ISAAC ASIMOV is un
foremost writer on sci
An Associate Professo
the Boston University School of Medicine, he has written two hundred books, as well as hundreds of articles in publications ranging from *Esquire* to Atomic Energy Commission pamphlets. Famed for his science fiction writing (his three-volume Hugo Award-winning THE FOUNDATION TRILOGY is available in individual Avon editions and as a one-volume Equinox edition), Dr. Asimov is equally acclaimed for such standards of science reportage as THE UNIVERSE, LIFE AND ENERGY, THE SOLAR SYSTEM AND BACK, ASIMOV'S BIOGRAPHICAL ENCYCLOPEDIA OF SCIENCE AND TECHNOLOGY, and ADDING A DIMENSION (all available in Avon editions). His non-science writings include the two-volume ASIMOV'S GUIDE TO SHAKESPEARE, ASIMOV'S ANNOTATED DON JUAN, and ASIMOV'S GUIDE TO THE BIBLE (available in a two-volume Avon edition). Born in Russia, Asimov came to this country with his parents at the age of three, and grew up in Brooklyn. In 1948 he received his Ph.D. in Chemistry at Columbia and then joined the faculty at Boston University, where he works today. His autobiography, IN MEMORY YET GREEN, will soon be available in an Avon paperback edition.

Other Avon Books by
Isaac Asimov

Adding a Dimension	36871	$1.50
Asimov on Physics	41848	2.50
Asimov's Biographical Encyclopedia of Science and Technology	44024	7.95
Asimov's Guide to the Bible, the New Testament	24786	4.95
Asimov's Guide to the Bible, the Old Testament	24794	4.95
Foundation	44065	1.95
Foundation and Empire	42689	1.95
Second Foundation	38125	1.75
The Foundation Trilogy	38026	5.95
From Earth to Heaven	42184	1.95
Life and Energy	31666	1.95
The Neutrino	38109	1.75
Of Time, Space, and Other Things	35584	1.75
The Planet that Wasn't	35709	1.75
Quasar, Quasar, Burning Bright	44610	2.25
The Solar System and Back	36012	1.50
The Universe: From Flat Earth to Quasar	42192	2.50
View From a Height	45336	1.25

LIFE AND TIME

ISAAC ASIMOV

AVON BOOKS
A division of
The Hearst Corporation
959 Eighth Avenue
New York, New York 10019

Published by arrangement with Doubleday & Company, Inc.
Library of Congress Catalog Card Number: 78-62644
ISBN: 0-380-47944-3

First Discus Printing, December, 1979

Printed in the U.S.A.

"The Gene Scene" appeared in the September 1977 issue of *The Saturday Evening Post,* copyright © 1977 by The Saturday Evening Post Company.

"Technology and Communication" originally appeared as "Not Enough Information, Too Much Information, and the Information Democracy" in *Dialogue in Technology #7* by Gould, Inc.

"One-to-One" originally appeared as "One-To-One-To-Wonderland" in the Autumn 1976 issue of *Creative Living,* copyright © 1976 by Northwestern Mutual Life.

"Farewell to Youth" appeared in the December 1974 issue of *The American Legion Magazine,* copyright © 1974 by The American Legion.

"Moving About" first appeared as "Earth Trek" in the November 1976 issue of *The Saturday Evening Post,* copyright © 1976 The Saturday Evening Post Company.

"The Ultimate Speed Limit" appeared in the July 8, 1972, issue of *Saturday Review of Science,* copyright © 1972 Saturday Review, Inc.

"The Coming Decades in Agriculture" is being published here for the first time.

"An Open Letter to the President" first appeared as "Isaac Asimov Advises the President" in the February 1977 issue of *Science Digest,* copyright © 1976 by The Hearst Corporation.

"Space and the Law" originally appeared as "Perspective of Space" in *American Law: the Third Century,* edited by Bernard Schwartz, copyright © 1976 by New York University.

"A Choice of Catastrophes" originally appeared as "20 Ways the World Could End" in the March 1977 issue of *Popular Mechanics,* copyright © 1977 by The Hearst Corporation.

TO

AL SARRANTONIO

VERONICA MIXON

AMY EINSIDLER

FOR ALWAYS BEING WILLING TO HELP
AND NEVER LOSING PATIENCE.

Contents

Introduction xi

PART ONE • LIFE PAST

1 Life 3
2 The March of the Phyla 18
3 Beyond the Phyla 36
4 The Gift of the Plants 57
5 The Brain Explosion 65
6 Man, the Overbalancer 69

PART TWO • LIFE PRESENT

7 The Myth of Less-Than-All 81
8 The Flaming God 88
9 Before Bacteria 94
10 The Face of the Moon 98
11 The Discovery of Argon 110
12 Water 124
13 Salt 132
14 Earth Shrugs Its Shoulders 141
15 Forget Me Not! 148
16 You Are a Catalog 158
17 The Gene Scene 169

PART THREE • LIFE FUTURE

18 Technology and Communication 179
19 One-to-One 193
20 Farewell to Youth 202
21 Moving About 216
22 The Ultimate Speed Limit 225
23 The Coming Decades in Agriculture 234
24 An Open Letter to the President 240
25 Space and the Law 246
26 A Choice of Catastrophes 259

Afterword 273

Introduction

Two and a half centuries ago, the English poet Alexander Pope, in An Essay on Man, *said: "The proper study of mankind is man."**

This would seem like advising us to retreat into a narrow parochialism; a human chauvinism.

Shall we do that? Shall we, ignoring all the vast universe, focus on ourselves alone, on our foibles and idiocies and microscopic grandeur and leave to one side all else? Surely such a sacrifice would be not only mean and selfish but would involve us in infinite loss.

But then we cannot quote without wrenching material out of context. Let us, therefore, take at least two lines—still out of context, but perhaps less dangerously so:

> *Know then thyself, presume not God to scan;*
> *The proper study of mankind is man.*

Those two lines set up Pope's antithesis between man and God; between a Universe running on natural law, on the one hand; and, on the other, by whatever it is that lies beyond the Universe and is bound by no limitations at all.

If we consider this division, we see that Science (with a capital S) precisely follows Pope's recommendation. It deals with the Universe and with the generalizations one can deduce and induce by observing the Universe, and by experimenting when that is possible. Whatever lies outside and beyond the Universe, whatever is not bound by law, whatever is not available for perception, observation, measurement, and experiment, is not the concern of Science. Such matters, Science presumes not to scan.

*Throughout this book when I use the word "man" I use it, as Pope did, in its general sense of "human being" including women—and children, too, for that matter.

Let me not imply that Science withdraws in humility. It does not necessarily turn its back on the Beyond, dazzled and worshipful, to busy itself with admittedly lowlier tasks.

When Napoleon looked through the volumes of Celestial Mechanics, *the monumental prerelativistic completion of Isaac Newton's work on gravitational theory, he said to its author, Pierre Simon de Laplace, "I do not see any mention of God in your description of the workings of the Universe."*

And Laplace answered grandly, "Sire, I have no need of that hypothesis."

But whether Science reacts to the Beyond with awe, patronization, or contempt, it leaves it to philosophers and theologians and it is, in my opinion, correct to do so.

Having said all this, however, there remains a great deal to the law-bound Universe that lies outside man. Ought we, then, to confine our studies to man alone?

If we think about it, such a study is not really confining, since man does not exist in a vacuum. Every other form of life impinges upon him, directly or indirectly; every inanimate environmental condition on Earth influences him; even such distant bodies as the Moon and the Sun exert an effect upon him. He is as subject to the laws of the Universe as is the smallest atom or the most distant quasar, and if it takes the study of the infinitely small, the infinitely large, the infinitely distant, or the infinitely abstract to elucidate those laws, then all those infinities are man's direct and selfish concern.

To study man, then, is to study the whole Universe.

Yet might it not distort our view of the Universe to look at it only through the peephole of its effect on us. Are we justified in the colossal conceit of judging everything by estimating its effect on us (like the Denver newspaper editor who insisted that a dogfight in Denver deserved more space in his columns than an earthquake in China).

After all, who but we ourselves cares about the effect of the Universe on ourselves?

The Earth existed for over three billion years with a load of life that did not include any hominids. The Earth, and the life upon it, got along well in that time, and would have continued to get along well (and, in some ways, better) if hominids had never appeared.

As for objects beyond Earth, none of them (except the Moon, recently and briefly) have been in any way affected by man, if we exclude the effect of unmanned probes and of feeble pulses of man-made electromagnetic radiation reaching them. The Universe, generally, does not know man exists and does not care.

We might argue, though, that man is absolutely a unique part of the Universe. It is a portion of the Universe that has, through a natural and exceedingly slow development, beginning at the Big Bang itself some 15 billion years ago, become sufficiently complex to be aware *of the Universe.*

We may not be the only portion of the Universe to have achieved this complexity. There may be billions of other species on other worlds circling other stars in this and other galaxies who face the Universe with awareness and curiosity. Some may have been in this presumably happy state for longer than our own species, may have evolved more sophisticated brains and developed more sophisticated observing and measuring instruments, so they may know and understand more than we.

Of these others, nevertheless, we have no evidence. However much we may feel certain that they must exist, that is only an inner assurance based on assumptions and deductions and shored up by no single direct observation.† *It remains conceivable that we may be the possessors of the only mind capable of observing the Universe.*

Well, then, if we cannot exist without the Universe, neither can the Universe be observed and understood without us. If we place Observed and Observer, or Riddle and Solution, on an equal basis, then man is as important as the Universe, and to study the Universe through man is legitimate.

In this collection of essays, I deal, more or less, with those aspects of the Universe that impinge quite directly upon man and other earthly life; past, present, and future. I call it, therefore, Life and Time.

† This, despite that awful farrago of silly nonsense, *Close Encounters of the Third Kind.*

PART ONE

LIFE PAST

I always make some attempt in these essay collections of mine to impose some sort of order on them. This is not easy to do since the essays were written at different times for different purposes and with absolutely no interconnecting order in mind.

I could impose a mechanical order, therefore, placing the essays in chronological order of publication—or in alphabetical order—or in order of decreasing (or increasing) length—or even in whatever order they come to hand.

I prefer, however, to make the order something more rational if I can; something that will lend meaning and make the book more than the sum of its parts.

In this case, I will try to arrange the essays so that they deal with the far past of life at the beginning and the far future of life at the ending, progressing regularly (or as regularly as I can manage, considering the miscellaneous nature of the essays) from past to future in the process.

But I don't intend to be slavish about it. I will begin, for instance, with an overview of life which I wrote for Collier's Encyclopedia *once.*

1

Life

One of the first ways in which we learn to classify objects is into two groups: 1. living and 2. nonliving.

In casual encounters with the material universe, we rarely feel any difficulty here, since we usually deal with things that are clearly alive, such as a dog or a rattlesnake; or with things that are clearly nonalive, such as a brick or a typewriter.

Nevertheless, the task of defining "life" is both difficult and subtle; something that at once becomes evident if we stop to think.

Consider a caterpillar crawling over a rock. The caterpillar is alive, but the rock is not; as you guess at once, since the caterpillar is moving and the rock is not. Yet what if the caterpillar were crawling over the trunk of a tree? The trunk isn't moving, yet it is as alive as the caterpillar. Or what if a drop of water were trickling down the trunk of the tree? The water in motion would *not* be alive, but the motionless tree trunk would be.

It would be expecting much of anyone to guess that an oyster were alive if he came across one (for the first time) with a closed shell. Could a glance at a clump of trees in midwinter, when all are standing leafless, easily distinguish those which are alive and will bear leaves in the spring from those which are dead and will not? Is it easy to tell a live seed from a dead seed, or either from a grain of sand?

For that matter, is it always easy to tell whether a man is merely unconscious or quite dead? Modern medical advances are making it a matter of importance to decide the moment of actual death, and that is not always easy.

Nevertheless, what we call "life" is sufficiently important to warrant an attempt at a definition. We can begin by listing some of the things that living things can do, and nonliving things cannot do, and see if we end up with a satisfactory distinction for this particular twofold division of the Universe.

1. A living thing shows the capacity for independent motion against a force. A drop of water trickles downward, but only because gravity is pulling at it; it isn't moving "of its own accord." A caterpillar, however, can crawl upward against the pull of gravity.

Living things that seem to be motionless overall, nevertheless move in part. An oyster may lie attached to its rock all its adult life, but it can open and close its shell. Furthermore, it sucks water into its organs and strains out food, so that there are parts of itself that move constantly. Plants, too, can move, turning their leaves to the sun, for instance; and there are continuous movements in the substance making it up.

2. A living thing can sense and it can respond adaptively. That is, it can become aware, somehow, of some alteration in its environment, and will then produce an alteration in itself that will allow it to continue to live as comfortably as possible. To give a simple example, you may see a rock coming toward you

and will quickly duck to avoid a collision of the rock with your head.

Analogously, plants can sense the presence of light and water and can respond by extending roots toward the water and stems toward the light. Even very primitive life forms, too small to see with the unaided eye, can sense the presence of food or of danger; and can respond in such a way as to increase their chances of meeting the first and of avoiding the second. (The response may not be a successful one; you may not duck quickly enough to avoid the rock—but it is the attempt that counts.)

3. A living thing metabolizes. By this we mean that it can eventually convert material from its environment into its own substance. The material may not be fit for use to begin with, so it must be broken apart, moistened, or otherwise treated. It may have to be subjected to chemical change so that large and complex chemical units (molecules) are converted into smaller, simpler ones. The simple molecules are then absorbed into the living structure; some are broken down in a process that liberates energy; the rest are built up into the complex components of the structure. Anything which is left over, or not usable, is then eliminated. The different phases of this process are sometimes given separate names: ingestion, digestion, absorption, assimilation, and excretion.

4. A living thing grows. As a result of the metabolic process, it can convert more and more of its environment into itself, becoming larger as a result.

5. A living thing reproduces. It can, by a variety of methods, produce new living things like itself.

Any object which possesses all these abilities would seem to be clearly alive; and any object which possesses none of them is clearly nonalive. Yet the situation is not at all clear-cut.

An adult human being no longer grows and many individuals never have children, but we still consider them alive even though they no longer grow and do not reproduce. Well, growth takes place at some time in life and the capacity for reproduction is potentially there.

A moth senses a flame and responds, but not adaptively; it flies into the flame and dies. Ah, but the response is ordinarily adaptive, for it is toward the light. The open flame is an exceptional condition.

A seed does not move, or seem to sense and respond—yet give it the proper conditions and it will suddenly begin to grow.

The germ of life is there, even though dormant.

On the other hand, crystals in solution grow, and new crystals form. A thermostat in a house senses temperature and responds adaptively by preventing that temperature from rising too high or falling too low.

Then there is fire, which may be considered as eating its fuel, breaking it down to simpler substances, converting it into its own flaming structure, and eliminating the ash which it can't use. The flame moves constantly and, as we know, it can easily grow and reproduce itself, sometimes with catastrophic results.

Yet none of these things are alive.

We must therefore look at the properties of life more deeply, and the key lies in something stated earlier: that a drop of water can only trickle downward in response to gravity, while a caterpillar can move upward against gravity.

There are two types of changes: one which involves an increase in a property called entropy by physicists, and one which involves a decrease in that property. Changes that increase entropy take place spontaneously; that is, they will "just happen by themselves." Examples are the downhill movement of a rock, the explosion of a mixture of hydrogen and oxygen to form water, the uncoiling of a spring, the rusting of iron.

Changes that decrease entropy do not take place spontaneously. They will occur only through the influx of energy from some source. Thus, a rock can be pushed uphill; water can be separated into hydrogen and oxygen again by an electric current; a spring can be tightened by muscular action, and iron rust can be smelted back to iron, given sufficient heat. (The entropy decrease is more than balanced by the entropy increase in the energy source, but that is beside the point here.)

In general, we are usually safe in supposing that any change which is produced against a resisting force, or any change that alters something relatively simple to something relatively complex, or that alters something relatively disorderly to something relatively orderly, decreases entropy, and that none of these changes will take place spontaneously.

Yet the actions most characteristic of living things tend to involve a decrease in entropy. Living motion is very often against the pull of gravity and of other resisting forces. Metabolism, on the whole, tends to build complex molecules out of simple ones.

This is all done at the expense of energy drawn from the food or, ultimately, from sunlight, and the total entropy change in the system including food or the sun is an increase. Nevertheless, the local change, involving the living creature directly, is an entropy decrease.

Crystal growth, on the other hand, is a purely spontaneous effect, involving entropy increase. It is no more a sign of life than is the motion of water trickling down a tree trunk. Similarly, all the chemical and physical changes in a fire involve entropy increase.

We become safer, then, if we define life as the property displayed by those objects which can—either actually or potentially, either in whole or in part—move, sense, and respond, metabolize, grow, and reproduce in such a way as to decrease its entropy store.

Since one sign of decreasing entropy is increasing organization (that is, an increasing number of component parts interrelated in increasingly complex fashion), it is not surprising that living objects generally are more highly organized than their nonliving surroundings. The substance making up even the most primitive life form is far more variegated and complexly interrelated than the substance making up even the most complicated mineral.

It may be that a simpler way of defining life would involve the finding of some sort of structure or component that is common to all living things and is absent in all nonliving things. At first glance, this might seem to be extremely difficult to do. Living things vary so widely in appearance that it is easy to suppose that though they may have certain abilities in common they do not have any structures in common.

Thus, though all living things can move, some do it by means of legs, others by fins, flippers, wings, ventral scales, cilia, flat immovable surfaces, and so on. The ability to move is held in common; but there is no one method of motion that seems to be held in common.

Indeed, the variety of life is such that much of the effort of the early biologists was expended in the classification of life forms: the attempt to place them all in an orderly system of groupings so that they might be studied with greater ease and to better advantage.

All visible forms of life, for instance, would seem to fall into one of two extremely broad groups: plants and animals.

Plants are rooted to the ground or float passively in the sea, while animals, on the other hand, frequently have the capacity for voluntary rapid motion. Plants have the capacity to use the energy of sunlight directly to power metabolism, making use of the green compound chlorophyll for the purpose. Animals lack chlorophyll and obtain their energy from the complex compounds of the food they eat. (Naturally, they eat plants, getting their energy from sunlight at one remove; or they eat other animals who have eaten plants and get their energy from sunlight at two or more removes.)

This division into plants and animals can even be extended into the microscopic world, for there are tiny organisms, invisible to the unaided eyes, which share key properties with the larger plants, or with the larger animals.

(Some argue, however, that the microscopic living things differ sufficiently from larger organisms, to warrant a separate, third division for themselves. Those who argue so call the microscopic organisms protists.)

The plant and animal kingdoms are each divided into finer divisions called phyla. The phyla are in turn divided into finer and finer divisions: first classes, then orders, families, genera, and, finally, species.

It is the species which represents a single kind of living thing. Man is a single species; the lion is another; the common daisy, another.

The number of different species, however, is enormous. There are about 400,000 different species of plants and about 900,000 different species of animals. (New species are constantly being discovered.)

What, then, can possibly be held in common by 1,300,000 species that differ as widely among themselves as do men and earthworms, whales and oysters, larks and moss, oaks and tadpoles, seaweed and elephants? (And this says nothing of many thousands, or even millions, of extinct species from trilobites to the giant moa.)

To the eye, there is no answer. Through use of the microscope, however, the answer was given long ago. In 1838, a German botanist, Matthias J. Schleiden, suggested that all plants were made up of separate microscopic units called *cells*. In 1839, a German zoologist, Theodor Schwann, extended the notion to animals.

Each cell is a self-contained unit, marked off from all the others by a membrane and capable of demonstrating in itself the various abilities associated with life. A cell, or parts of it, can move, sense, and respond, metabolize, grow, and reproduce.

Organisms large enough to see with the naked eye are made up of large numbers of cells. An adult human being contains some fifty trillion (50,000,000,000,000). Each cell in such a multicellular organism is so adapted to the presence of others as no longer to be capable of living in isolation. Nevertheless, there are some cells that are indeed capable of living independently. Most forms of microscopic creatures are made up of single cells; they are unicellular organisms. And even large creatures begin life as single cells. Each human being got his start as a fertilized ovum—one cell.

Moreover, though organisms may differ enormously, the microscopic cells of which they are composed do not differ nearly as much. A whale cell is much more like a mouse cell than a whale is like a mouse.

All plants and animals are made up of cells, and those parts of a living organism that are not made up of active cells are not alive. (A tree's bark is not alive, nor is an animal's hair, nor a bird's feathers, nor an oyster's shell—which does not mean the organism can necessarily live without that nonliving portion.) Furthermore, no nonliving thing is made up of active cells—though a freshly dead organism is made up of dead cells. (Some cells may live on briefly after the overall death of the creature; but before long, all are dead.)

The phrase "active cells" implies that the cells can perform the actions characteristic of life, so we are now defining life as the property of objects made up of cells possessing the ability to move independently, sense and respond adaptively, metabolize, grow, and reproduce.

This eliminates any possibility of imagining such noncellular objects as crystals and fire as having life.

Yet there still remains a source of confusion. In 1892, a Russian bacteriologist, Dmitri Ivanovsky, discovered a disease agent so small it could easily pass through a filter designed to bar the passage of even the smallest bacterium. Thus were discovered the viruses which are much smaller than cells and which, in isolation, show none of the ordinary criteria of life. Indeed, they can even be crystallized and, at the time this was discovered, crystallization was felt to be a property that could not possibly be associated with anything but nonliving chemicals.

Yet once in contact with cells, individual virus particles can somehow penetrate the cell membrane, bring about specific metabolic reactions and reproduce themselves. In some ways and under special conditions, they show unmistakable properties associated with life. Are viruses, then, alive or not?

If life is defined in terms of cells, viruses are not alive, for they are much smaller than cells. But can life be defined still more fundamentally and usefully so as to include viruses as well? To see if that is so, let's consider the substances of which cells are composed.

Cells contain an enormously complex mixture of substances, but these are built up out of only a few elements. Almost all the atoms they contain are of half a dozen different kinds: carbon, oxygen, hydrogen, nitrogen, phosphorus, and sulfur. There are smaller quantities of other atoms such as those of iron, calcium, magnesium, sodium, potassium, and traces of copper, cobalt, zinc, manganese, and molybdenum. There is nothing in these elements themselves, however, that gives any clue to the nature of life. They are common enough in nonliving objects.

The atoms within the cell are grouped into molecules that, in the main, fall into three types: carbohydrates, lipids, and proteins. Of these, the protein molecules are by far the most complex. Whereas molecules of carbohydrates and lipids are usually made up of atoms of carbon, hydrogen, and oxygen only, proteins invariably include nitrogen and sulfur atoms as well. Whereas carbohydrate and lipid molecules can be broken down to simple units of two to four kinds, the protein molecule can be broken down into simple units (amino acids) of no less than twenty different varieties.

Proteins are of particular importance in connection with the thousands of different chemical reactions constantly proceeding within cells. The velocity of each different reaction is controlled by a class of protein molecules called enzymes—a different enzyme for each reaction. The cell contains a large number of different enzymes, each present in certain amounts and, often, in certain positions within the cell. The enzyme pattern determines the pattern of chemical reactions and therefore controls the nature of the cell and the traits of the organism built up out of the cells.

The properties of the enzymes molecule depend on the particular amino-acid arrangement it possesses. The number of possible arrangements is inconceivably large. If a molecule is made up of 500 amino acids of 20 different kinds (average for a protein), the total number of possible arrangements can be as high as 10^{1100} (a figure we can write as a 1 followed by 1,100 zeroes). How, then, does the cell form the particular arrangement needed for particular enzymes out of all those possibilities?

The answer to this question seems to lie in the chromosomes, small threadlike structures in a small body called the nucleus, usually located near the midpoint of the cell. Whenever the cell is in the process of division, each chromosome forms another just like itself (replication). Each of the two daughter cells formed at the end of the division gets its own duplicate set of chromosomes.

Chromosomes are made up of protein in association with an even more complex molecule called deoxyribonucleic acid, usually abbreviated as DNA. The DNA contains within its own structure the "information" needed for the construction of specific enzymes and also for the replication of itself so that it can continue to guide the construction of specific enzymes in the daughter cells. Every single creature has the DNA molecules to build its own enzymes and no other.

Is it possible that, just as certain organisms may consist of individual cells, still simpler ones may consist of individual chromosomes? Apparently yes, for viruses are very much like individual and independent chromosomes.

Each virus is composed of an outer coat of protein and an inner molecule of DNA (or, in some cases, a similar molecule, ribonucleicacid or RNA). The DNA or RNA manages to get inside a cell and there supervises the production of enzymes designed to produce more virus molecules of exactly the type that invaded the cell.

If, then, we define life as that property possessed by objects containing at least one active DNA, or RNA, molecule, we may have all we need. The cells of all plants and animals, and of all unicellular organisms, and the molecules of all viruses as well, all contain at least one DNA or RNA molecule (and, in the case of the cells, many thousands). As long as these molecules are capable of guiding the formation of enzymes, the organism is alive with all the attributes of life. Objects that were never alive, or were once alive but are no longer so, do not possess active molecules of DNA or RNA.

* * *

Living creatures represent different levels of complexity and organization. A large creature is generally more complex than a small one of the same kind, if only because it has more interrelated parts. Animals in general are more complex than plants. For one thing, animals have particularly complex tissues such as muscle and nerve, which plants lack. A mouse may fairly be considered to be more complex than an oak tree for that reason.

The most complex structures found in the animal organism generally are the brains; and these are most complex in certain mammals. The largest brains of all are those of men, elephants, and whales. The human brain, for instance, weighs three pounds and is composed of ten billion nerve cells and ninety billion auxiliary cells, with each of the nerve cells connected to perhaps a thousand others, and with each individual nerve cell enormously complex in itself. Pending further information concerning the complexity of the brains of elephants and whales, it seems fair to say that the human brain is the most highly organized object known to us.

Naturally, this level of organization was not achieved in a bound, but was the product of at least three billion years of slow changes. The changes themselves were produced by random imperfections in DNA replications, which led to corresponding changes in enzyme structure and therefore to the reaction pattern in cells. Those particular changes survived which, for one reason or another, proved beneficial to the organism under the particular conditions existing about it. (Such a theory of evolution by natural selection was first advanced by the English biologist, Charles Darwin, in 1859).

But how did the whole process get started? Right now, every cell is formed from a previously existing cell. Every DNA molecule is produced by a previously existing DNA molecule. Yet surely life did not always exist, since there was a time when Earth itself did not exist. How, then, did the first cell, the first DNA molecules, come into existence?

Many suppose that some supernatural being created life. Scientists, however, prefer not to seek explanations in the supernatural. They suppose, rather, that the known laws of physics and chemistry suffice to offer possible mechanisms for the origins of life.

Could life have come from some other world? The most

popular version of that theory was advanced in 1908, when a Swedish chemist, Svante Arrhenius, suggested that living spores might be driven across the great gaps of space between the stars by the pressure of starlight. Some would have fallen on the youthful Earth and would have begun life there. But that only postpones the difficulty. How did life originate on the planet from which the spores came?

In recent years, scientists have begun to consider the chemical makeup of the Universe generally. The Universe is believed to be about 90 per cent hydrogen. When the Earth was formed, its atmosphere must therefore have been rich in hydrogen-containing compounds. If we consider hydrogen combining with other common elements, we can imagine Earth's atmosphere to have consisted originally of methane (hydrogen combined with carbon), ammonia (hydrogen combined with nitrogen), and water (hydrogen combined with oxygen).

What would happen if such compounds and others like them were exposed to a bath of energy from the sun. As they absorbed the energy, would they build up more complicated compounds?

In 1952, an American chemist, Stanley Lloyd Miller, prepared a mixture of chemicals such as were believed to exist on the early Earth. He subjected them to the energy of an electric discharge for a week, then he analyzed his mixture. He found that more complicated compounds had indeed been formed. In particular, two or three of the simpler amino acids that go into the makeup of proteins were formed.

Ever since then, similar experiments have been performed by many groups, and it has turned out that the basic compounds associated with life can in this way be formed from the very simple compounds that were probably found on the early Earth.

The American chemist, Sidney W. Fox, began with amino acids and subjected them to heat. He found that proteinlike molecules were formed which, on the addition of water, clung together in little microspheres about the size of small bacteria. Can cells have made their primitive beginning in this way?

It may have taken a billion years or so for compounds to become complex enough, and cells to grow complicated enough, to form objects we might recognize as very simple forms of life. Once that happened, the living cells would compete with each other for food, and those that were more efficient would survive at the expense of others. With time, cells would grow more and more organized and complex.

Originally, the cells would have to use as food the complex compounds built up by the slow action of the Sun's ultraviolet radiation. In the process, the methane and ammonia in the atmosphere would be changed to carbon dioxide and nitrogen.

Eventually, certain cells developed the use of chlorophyll, which allowed them to use the Sun's visible light as a source of energy, in a process called photosynthesis. This enabled them to build up complex molecules far more rapidly.

In photosynthesis, carbon dioxide is consumed and oxygen is liberated as a waste product. Eventually, the carbon dioxide and nitrogen atmosphere would be converted to the oxygen and nitrogen atmosphere we have today.

Is it possible that life started only once and that from the initial life form, all present life developed? That would be why, perhaps, all species have a basic chemical similarity. Or did it begin a number of times, with each life form basically similar to all the others because only one form of chemistry can lead to substances complex enough to demonstrate the properties of life?

It is impossible to check on this by watching for life to form under our nose on Earth as once it did. Billions of years ago, life had a chance to form because there was no life already existing. Nowadays, any complicated molecule forming on the way to life would quickly be eaten by some already existing life form.

But what of other planets? Ordinarily, we do not think of the other planets of the Solar system as capable of supporting life. Earth life is adapted to earthly conditions so that most forms of life require oxygen and water, a moderate temperature, the absence of poisonous compounds, gravity and air pressure not too different from what actually exists, and so on.

The Moon, then, would seem no fit abode for our kind of life because it lacks air and water. Mars's thin atmosphere has no oxygen and it possesses very little water.

Nevertheless, even though men and other highly organized creatures could not live unaided on the Moon or on Mars, it is possible that simple protistlike creatures might have developed. Underneath the Moon's outer surface, there are mild temperatures where small quantities of trapped water and gases may exist. There, a thin population of bacteria might live. On Mars, there is even the possibility of simple lichenlike plants.

If these extraterrestrial life forms actually existed, and were like our own, chemically, that would be strong evidence in favor of only one possible chemical basis of life. If they were not like our own, how fascinating to study a second (and third?) chemical basis of life we can't even conceive now.

No wonder space scientists are anxious to sterilize all man-made objects that will touch down on an alien world. If we contaminate such a world with our own bacteria, the most exciting experiments in biological history may be robbed of their meaning.

But what about highly developed life? What about intelligence?

There doesn't seem to be any world in our Solar system that could support highly developed life based on earthly life chemistry. For that, we would have to look to planets circling other stars.

There, the possibilities seem good. There are, in our own Galaxy alone, about 135,000,000,000 stars. According to modern theories of planet formation, almost all ought to have a system of planets. Some of the stars ought to be rather like our Sun, and some of these ought to have at least one planet like the Earth at the proper distance.

In 1964, the American astronomer, Stephen H. Dole, taking into account as much information as possible, estimated that the number of planets like Earth in our own Galaxy alone, would be 645,000,000. (And there may be as many as a hundred billion other galaxies in existence, too.)

On any planet very much like our Earth, chemical changes would take place similar to those that took place here. Life would form—but even if it formed on the same chemical basis, no one could tell how it would appear structurally. Considering in how many different ways life developed on Earth and how many hundreds of thousands of different species formed, it seems unlikely that a similar wild variety would not form there, and it would be an almost impossible chance to have a species there closely resemble some species here.

Yet some alien life forms might develop intelligence and that intelligence, at least, might resemble ours. Unfortunately, there is no way of estimating the chances of the development of intelligence.

Still, even if intelligence eventually developed only one time in every million life-bearing planets, there might be over six hundred different types of intelligent beings in our own Galaxy alone.

Unfortunately, the Universe is vast. Our own Galaxy is so huge that if even as many as 645,000,000 planets were spread out evenly, the nearest one to us would be some two dozen light-years distant, and the nearest intelligence (even assuming these existed) might be no closer than 25,000 light-years.

Whether such distances can ever be spanned or not is uncertain. Perhaps the various intelligences are forever insulated from each other, or perhaps if some of them are more advanced than we are, then they may come visit us someday (when we are ready for them) and invite us to enter a United Galaxy Organization.

What about life forms radically different from ours, based on altogether different kinds of chemistry, living in completely hostile (to us) environments? Could there conceivably be a silicon-based life, in place of our own carbon-based one, on a hot planet like Mercury? Could there be an ammonia-based life, in place of our own water-based one, on a cold planet like Jupiter?

We can only speculate. There is absolutely no way to tell at present.

We can wonder, though, whether human astronauts, exploring a completely alien planet, would be sure of recognizing life if they found it. What if the structure were so different, the properties so bizarre, that they would fail to realize they were facing something sufficiently complex and organized to be called living?

For that matter, we may be facing such a necessary broadening of the definition right here on Earth in the near future. For some time now, men have been building machines that can more and more closely imitate the action of living things. These include not merely objects that can imitate physical manipulations (as when electric eyes see us coming and open a door for us) but also objects that can imitate men's mental activities. We have computers that do more than merely compute; they translate Russian, play chess, and compose music.

Will there come a point when machines will be complex

enough and flexible enough to reproduce the properties of life so extensively that it will become necessary to wonder if they are alive?

If so, we will have to bow to the facts. We will have to ignore cells and DNA and ask only: What can this thing do? And if it can play the role of life, we will have to call it living.

This essay and the one following are the oldest in this collection. They have managed, for twenty years, to escape my generally relentless determination to see to it that everything I have written (within reasonable limits) receives the relative permanence of book publication. Let me explain how that came about.

In the 1950s, I wrote occasional science essays for Astounding Science Fiction. *A dozen of these articles were put together in my first essay collection,* Only a Trillion *(Abelard-Schuman, 1957).*

But then, in 1959, The Magazine of Fantasy and Science Fiction (F & SF) *asked me to do a monthly science column for them. I accepted gladly and have been turning them out for twenty years now without missing an issue. What's more, the good people at Doubleday & Company, have been equally assiduous in publishing collections of these* F & SF *essays at seventeen-month intervals, on the average.*

My F & SF *essays occupied my attention so fully that I gave no thought to my earlier essays in* Astounding. *Those few essays that were written for* Astounding, *after the appearance of* Only a Trillion *and before my stint in* F & SF, *went uncollected by me. Some, of course, are by now out-of-date, but this one, and the next, are, in my opinion, still precisely on target. I am delighted to be rescuing them from oblivion now.*

2

THE MARCH OF THE PHYLA

Life, presumably, began with a single nucleoprotein molecule—which is equivalent, today, to a gene within a cell, or to a small virus outside one. It progressed next to an association of nucleoprotein molecules—equivalent today to a chromosome within a cell or a large virus outside one.

The advantage of the molecular association was that the weakness of one molecule of the group might be compensated for by the strength of another. In this way, specialization was possible. Each molecule in the group might be unviable singly because of some essential lack, but each might function far above average in another respect. A skillful combination in which no essential lack ran through all members of the group might result in an organism which together functioned far better than any collection of all-round-good-but-nothing-special individual molecules.

A second change was the conversion of the bare group of nucleoprotein molecules to one that was surrounded by food stores and useful chemicals, all held together by a membrane that could control the nature and quantity of the substances entering and leaving. The "virus" had become a "cell." Presumably, the first cells were simple cells, with low levels of organization, equivalent to the bacteria and simpler molds of today.

Now it is usual to consider the change from virus to cell an "advance"—a climb up the tree of life, so to speak. But what do we mean by that? What makes one organism "higher" or "more advanced" than another?

Is it the mere test of survival? If so, the question of virus vs. cell boils down to "no decision." Both viruses and cells exist to this day and neither is likely to be wiped out by anything short of planetary cataclysm. As a matter of fact, viruses are rather harder to kill than cells so that perhaps the move from virus to cell was a retreat rather than an advance. In fact, maybe the development of life in general was a retreat, since a rock or a molecule of water will withstand changes that will kill even a virus.

But words can be defined arbitrarily. We are human beings looking at the universe through human senses and interpreting the messages we receive by means of a human brain swayed by human emotion. It is, therefore, perfectly natural to define "advancement" in human terms.

A human being "advances" when he rises in the social scale by use of wealth, intelligence, force or any other means. The measure of his advancement is his ability to control his environment or his freedom from the pressures of his environment. (What man does not wish to "be his own boss"; and what is that but a way of longing for fewer pressures and greater control.)

Applying this anthropomorphic outlook to life generally, we can say that the more an organism is in control of its immediate environment or the more it is free of its pressures, the more advanced it is.

Let's take an example. A virus has the wherewithal to organize a food supply into duplicates of itself, but it must take the food supply that comes its way. If the necessary molecules bump into it, wonderfully well. Otherwise, it must wait.

The cell has the capacity to store molecules that serve as food. During a lucky period of food density, it can hang on to more than it can use at the moment—which the virus cannot do—and save it for future use.

Thus the cell has freed itself, to a certain extent, of one of the elements of chance in its environment. It is less immediately dependent than is the virus upon its environment for food.

Again, cells have the capacity for movement at will; viruses do not. This does not mean that *all* cells move. It does mean that some do; the potential is there. However, no virus moves freely and no virus ever did as far as we know; the potential is simply not there.

A virus must depend upon some external force—such as a current of water—to move it against a food supply, or a food supply against it; or to move it away from a danger, or the danger away from it. The moving cell, however, can actively search for food. It can and does develop chemical devices to detect food (or danger) at a distance. Such detection can set up a chain of automatic changes that result in motion toward the food or away from danger.

Again the cell is less the slave of its environment than the virus is. By that measure, the cell is the more advanced.

An organism which is in greater control of its environment than a certain competitor is bound to win out in the competition. When cells and viruses compete for the same food supply, the cell can go after the food and grab it while the virus must wait for the food to come to it by chance. The cell can take all it can get and store the surplus; the virus must take only what it needs and let the rest go.

As a result, these are the possibilities open to the virus: First, it can simply be beaten in the competition and cease to exist. Second, it can retreat from the competition and find a place for itself where cells do not exist. Third, it can follow the old adage to the effect that if you can't lick them, you must join them, and become a parasite.

Those viruses that exist today have adopted the third path. If there were ever free-living viruses, there are none now.

The viruses of today use cells as their food supply and survive beautifully as a result. The cell utilizes its greater control of the environment to build up a food supply and then the virus steps in and makes use of the food supply.

This is so attractive a way out of a disadvantageous competition that, as an alternative, it has been chosen time and time again in the course of evolution. Some types of organisms, to be sure, ended in extinction. Some were forced into less desirable living niches where there was less competition, but preserved their independence and, in some cases, made startling advances in unexpected ways.

But always there was the lure of parasitism. There are parasites at every level of advancement of life; and, by and large, if mere survival is counted, parasitism has proven brilliantly successful.

But the parasitic control of the environment is a regressive one. It works by picking an extremely specialized environment and tying one's self to it completely. A minor alteration in the environment—such as death of the host organism—kills the parasite. Furthermore, in adjusting to the environment, there is an inevitable regression to lower levels of organization. The environment after all is so ideal that it makes practically no demands on the parasite. So the parasite makes its advances only along the pathways of retreat.

Parasitism is a very good life; a Garden of Eden.

It is to be avoided like death.

As cells grew more elaborate in their race for greater control of the environment and for consequent advantage in their eternal mutual competition for food and safety, a fundamental split in variety took place which persists to this day.

Some cells developed chlorophyll and were freed of the struggle for food in the sense that they thence-forward needed only water, carbon dioxide, certain minerals and sunlight—all of which were virtually ubiquitous and inexhaustible. These cells and their descendants are the members of the plant kingdom.

The remaining cells, which, with their descendants, make up

the animal kingdom, get along without chlorophyll. To do so, they must eat ready-made organic matter; either the remains of once-living cells, or the intact plant cell, or an intact animal cell that has been living on one or both of the first two items.

In a sense, then, animal cells live on carbon dioxide, water, minerals and sunlight, too—but with at least one middleman involved. Isn't this a form of parasitism on the middleman? Isn't this the kind of death-in-the-garden-of-Eden I have just warned against?

The evidence in favor of viewing animal life, generally as parasitic is this: plant life, in some of its forms, can continue to exist indefinitely, even though all animal life were destroyed, but the reverse is not true. No animal life would exist for more than a short period after the destruction of plant life.

Furthermore, since animal life lives on solar energy via a middleman, there is the natural wastage associated with middlemen everywhere. It takes roughly ten pounds of plants to support one pound of animal, so that the total mass of living matter on Earth is ninety percent plant and ten percent animal.

And yet what about the arguments on the other side. Animal life does not fulfill the main qualification of parasitism; that its food become its environment. A true parasite lives within its food and need not seek it—except for the original seeking that establishes it within its host. Animal life must seeks its food constantly and is, therefore, no true parasite. The fact that its particular foodstuff is a plant cell rather than, say, a small pebble is but a difference in detail.

In fact, it is plant life that is surrounded by the air, water, minerals and sunlight that are its food; and, therefore, it is the plant cell that is the true parasite. This is not the usual way of looking at it, I know—in fact, as far as I know, this is original with me—but consider that the plant cell shows some of the marks of the parasite.

It displays a decreased control of its environment as compared with the apparently simpler bacteria. Some bacterial cells can move at will; plant cells cannot. Plant cells are as motionless as viruses. Plant cells store and expend energy slowly and live at a low level of intensity. In fact, they don't "live," they "vegetate."

The animal cell, on the other hand, can expend energy at a rate limited only by the amount of plant material it can eat and metabolize per unit time. By the ability to move at will and to live more quickly in general, the animal cell can control its environment far more than the plant cell can. (To put it in its simplest terms: You can bite a carrot but the carrot can't bite you back.)

The conclusion, then, is that the animal cell is more advanced than the plant cell.

In general, continued elaboration of cells almost inevitably involves increases in size. The more complex cells are the larger ones. The larger a cell, the longer the chromosomes it can hold, or the more numerous; the more enzymes it can contain; the more food it can store; the more energy it can generate; the more it can divide itself into specialized subdivisions. In short, a large cell can do more than a small cell and is likely to be, by the definition being used here, more advanced.

But as cells grow larger, trouble arises. The rate at which food enters a cell, and wastes depart, depends on the surface area of the cell. The total food requirements of a cell depend on its volume. But as a cell increases in size, the volume increases as the cube of the diameter, the surface only as the square. If a spherical shape is maintained, a size is quickly reached in which there is no longer enough surface to feed the increased bulk.

An alternative would be to abandon the spherical shape. Cells might be long or flat or irregular. The only trouble is that the spherical shape is the one which requires the least energy to maintain. Any departure involves an input of energy, an input which increases with the size of the cell. Small bacterial cells may be rod-shaped, but for larger cells in isolation, this is a major feat. The amoeba may thrust out blunt pseudopods, the paramecium may be slipper-shaped, but even so maximum size is quickly reached.

Another alternative is for cells to remain small and reasonably spherical, but stick together after cell division. In this way, a group of cells are formed which have whatever advantage sheer mass brings, while leaving each individual to be within the safety limit of the "square cube" law.

Thus, cell colonies, both plant and animal, can be and have been formed. The advantage of a cell colony, if it is simply a collection of completely independent cells and nothing more, over so many separate cells, is not great. However, the existence of a cell colony makes possible specialization at the cellular level.

The most successful cell colonies in the animal kingdom, for instance, are the sponges, which can grow to enormous sizes when compared with individual cells. Sponges are made up of several types of specialized cells, each of which performs a certain job particularly well.

There is a type that secretes a gelatinous fibrous material that both supports and protects a colony, so that the colony as a whole is safer and better protected from the stresses of the environment than any individual cell can be. Other sponge cells have flagella that can whip up a current, which will carry food particles into the colony and wastes out. Still others contain pores through which the current will pass.

What it amounts to is a division of labor, with a consequent overall increase in efficiency.

Yet in a cell colony, even so complicated a one as the sponges, the individual cell has not given up its birthright. Any one cell of a sponge can and sometimes does, wander off on its own and start a new colony.

But let's extend this trend to its logical conclusion. To increase the efficiency of a cell colony, more and ever more specialization would be required. Each cell must get better and better at its particular task even if it means that other abilities are allowed to grow vestigial. The deficiencies of one cell will, after all, be made up for by its neighbors. (This is the conversion of gene to chromosome on a higher level.)

Eventually, the individual cell of a colony becomes so specialized that it can no longer exist on its own; only as part of a group.

When this point is reached, we are dealing with more than a cell colony. We have a "multicellular organism."*

But now the individual cell is completely at the mercy of the multicellular organism as a whole. The cell cannot live outside the organism and is, therefore, a parasite upon the organism. Is not this a regression?

If you concentrate on the individual cell, it is. But the cell is no longer the organism. It no longer counts as a measure of "advancedness"; it is the entire cell-collection now that has the "consciousness of life."

*In the less advanced multicellular organisms, relatively small groups of cells from the organism can, if torn loose, survive and serve as the nucleus of a new organism. This is "regeneration." As multicellular organisms advance through ever-increasing specialization, powers of regeneration grow progressively less.

We can see that in ourselves. It doesn't matter to us that millions of our red blood cells die each minute, or that our skin is constantly renewed only by the continual dying of cells just below the epidermis. A wound damaging or killing millions of cells is of no permanent consequence provided only that it heals. If absolutely necessary, we will sacrifice a leg to save a life. In short, while the consciousness of the whole persists, the parts are but of secondary consequence.

We have no choice but to apply this principle to other multicellular organisms even when we're pretty sure that "consciousness of life" in the human sense does not exist among them. The equivalent, whatever it is, does, so with the advent of the multicellular organism, we must consider the organism only, not the cells composing it.

I should mention here that what I am calling "advance" does not necessarily imply advantages only. The cell is more advanced than the virus but is easier to kill. Although the cell has greater control of its environment within certain limits, it can less well withstand the stress of the environment beyond those limits.

Similarly, a multicellular organism is, in some ways, more susceptible to death than an individual cell is.

An individual cell is potentially immortal. Given sufficient food and safety, it will grow and divide forever. The multicellular organism, however, depends not only on the cells that compose it but on the organization among them. All its cells but for an insignificant few might be in working order. Nevertheless, if the malfunction of the few destroys intercellular organization, death must ensue for the entire organism and for all the healthy cells that compose it.

Intercellular organization, moreover, is never everlasting. A multicellular organism, though living with ample food and in complete safety, must nevertheless eventually die.

Nevertheless, advantages and disadvantages must be weighed against each other. Looking back along the winding path of evolution, we must conclude that the greater flexibility of the cell within limits more than made up for its greater fragility outside those limits. Similarly, the greater flexibility of the multicellular organism more than made up for the fact that inevitable death came into the world.

In fact, even an apparent disadvantage could be made into a consummate victory. To avoid extinction of the species, provision must be made for the formation of one or more new multicellular organisms before the old one died. This was done and eventually the system was refined to the point where it required the cooperation of two organisms to produce a new one. With the invention of sexual reproduction, there came the eternal shuffling of chromosomes with each generation. Variation among individuals became more common and more drastic and the course of evolution was hastened.

It is interesting to note that the plant kingdom, with its easier life and its parasitism on sun, air and water, made this advance into multicellularity neither as extensively nor as intensively as the animal kingdom. In fact, the plants of the sea never advanced beyond the cell-colony stage. The most elaborate seaweed is only a cell colony.

It is only when plants invaded dry land and water and minerals became less easy to get that specialized organs had to be developed to collect them from the ground, others to collect light from the sun, and others to communicate water from below and food from above to other parts of the organism. Even so, the most elaborate tree is not as elaborate as even a simple animal. No plant, for instance, has a nervous system, or muscles, or a circulating blood system. No plant can move freely in the sense that an animal can.

All the types of organisms I have so far mentioned, still survive in today's world after possibly two billion years of environmental vicissitudes, though not necessarily in their original form. All will undoubtedly continue to survive, barring planetary cataclysm.

However, mere survival is nothing. On the basis of control of environment, the types of organisms can be presented as in Figure 1. The arrows included are *not* intended to indicate lines of descent, of course. They indicate instead the direction of increasing control of environment. It seems not a hard decision to make; obviously the multicellular animal organism is the most advanced of those listed in the figure. We might say it "rules the Earth."

Multicellular animals, to which I will now confine myself, are divided into a number of broad groups called "phyla"—singular, "phylum." Within each phylum there may be wide diversity, but there is retained a certain uniformity of general body plan.

For instance, you may not think there is much similarity between yourself and a fish, but both you and a fish have bones arranged in similar fashion; you both have a heart; you both have blood containing similar chemicals; you both have four

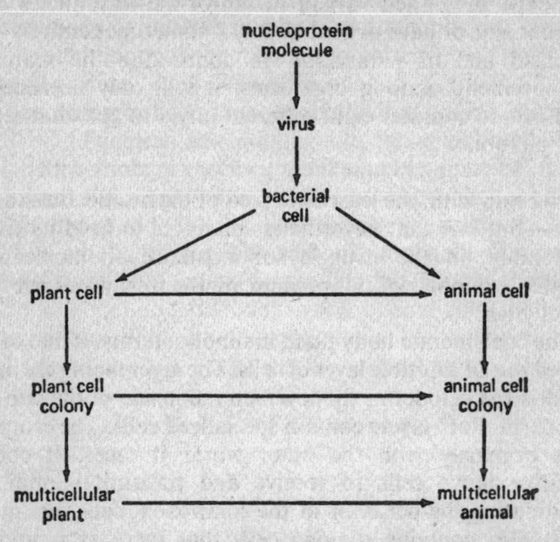

Figure 1

limbs arranged in pairs; you both have a pair of eyes and a mouth forming part of the head and so on. Anatomists and zoologists would find hundreds of other gross physical similarities.

The point is that you and the fish belong to the same phylum.

Now compare yourself with an oyster. You may be at a loss to find similarities except for the obvious one that you and the oyster are both multicellular. Different phyla, you see!

Of course, the exact division into phyla is a man-made thing and not all authorities agree on just what creatures go into which phyla. (Nature somehow never did organize itself with the future convenience of human classifiers clearly in mind. Sad, but true.)

Nevertheless, Van Nostrand's *Scientific Encyclopedia,* which I happen to have handy, lists twenty-one phyla of multicellular animals.

Interestingly enough, all twenty-one attempts at varying the basic organization worked, in the sense that creatures belonging to each phylum survive today and will probably go on surviving into the foreseeable future. There are no fossil records of any distinct phylum—as far as I know—that is now entirely extinct.

Over half the phyla, however, although surviving, have been

distinctly beaten by the competing phyla. These beaten ones now exist in limited variety in out-of-the-way niches of the environment or have drifted largely—sometimes entirely—into the dead end of parasitism. In continuing the search for "advancement" among organisms, it will only be necessary, therefore, to consider eight different phyla to get what seems a clear picture.

To begin with, the least advanced of the multicellular animal phyla—but one that nevertheless manages to hold its end up, respectably, in the struggle for existence—is Coelenterata. Familiar examples of this phylum are the freshwater hydra and the jellyfish.

The coelenterate body plan, in simplest terms, is that of a cup shaped out of a double layer of cells. The layer facing the outside world is the ectoderm; the layer on the inside of the cup is the endoderm. Both layers contain specialized cells. The ecotoderm deals primarily with the outer world it faces. It contains primitive nerve cells to receive and transmit stimuli, thus coördinating the behavior of the component cells that make it up. It also contains stinging cells that serve as weapons of offense, capturing smaller organisms. The endoderm, on the other hand, is a food-centered layer. It contains cells specialized to secrete juice that digests the captured organisms that prepare them for absorption.

A particular advance made by the coelenterates is the possession of the interior of the cup as a private bit of the ocean. In cells and cell colonies, however complicated, food particles must be engulfed into the body of a cell before it can be used.

The coelenterates can, instead, pop food particles into the interior of the cup—which is a primitive digestive sac, or "gut"—and there digest it. The cells of the endoderm need only absorb the dissolved products of digestion, not the particle itself. In this way, many food particles can be handled at once; and individual food particles considerably larger than a cell can be handled. Any improvement in the feeding plan automatically means an important improvement in the control of the environment, so that coelenterate, although the lowest of the multicellular organisms, is much advanced over even the most specialized of the cells or cell colonies.

Another phylum, Platyhelminthes, has added further refinements to the coelenterate body plan. (This phylum, which may also be referred to as the "flatworms," contain well-known parasitic forms, notably the various tapeworms. It also contains free-living forms, the best known of which is a little half-inch creature called "planaria.")

For one thing, the flatworms possess a third layer of cells, called the mesoderm in the space—"coelom"—between the ectoderm and the endoderm. (And that ends it. No fourth layer has ever been developed in any phylum.) The mesoderm is not primarily concerned with relations with the outer world, as is the ectoderm; nor with feeding, as is the endoderm. Instead the mesoderm can be used to form organs that the body requires for internal specialization. (The usefulness of this invention is proved by the fact that no phylum after the flatworms has ever abandoned it.)

For instance, the flatworms use the mesoderm to form contractile fibers that are the first animal muscles. They also form special reproductive organs and the beginnings of excretory organs. All these introduce new specialization and hence new and more efficient ways of responding to the environment. Muscles, as an example, enable the flatworms to move with greater ease and efficiency than do the coelenterates.

In addition, the flatworms display bilateral symmetry. This means that the right and left halves are mirror images, but the front and rear ends are not. The flatworms have a differentiated "head" and "tail" and it is the head which is generally pointed in the direction of movement.

In single-celled creatures, in cell colonies and in the coelenterates, there is radial symmetry. These creatures must be equally on guard on all sides. In flatworms, since it is the head which is particularly in advance and entering the unknown, it is the head which needs to be particularly sensitive to stimuli. Concentrating the area of response to stimuli means increasing the efficiency of the response and thus allowing for better potential control of the environment.

As an example, the flatworms have developed the primitive nerve cells of the coelenterates into an organized nerve network with a concentration in the head area where it is most needed. The flatworms, in other words, have invented the first primitive brain.

However, both coelenterates and flatworms still depend for nourishment on simple absorption of food from the outside world into the various component cells. This prevents them from ever attaining a great bulk—with the advantage of an increased potential efficiency—since each cell must remain within a certain distance of the outside world, or not enough food and oxygen will reach them.

To be sure, there are giant jellyfish, but their long stingers are very thin and their voluminous "bell" is composed mostly of a very watery gelatinous material—hence "jellyfish"—with the actual living cells very near the surface. There are also giant flatworms—such as seventy-foot tapeworms—but these are flat as tape, as the names imply. They can never be very thick.

To enable a multicellular organism to achieve real bulk, as distinguished from simple length, a new invention was needed. That was supplied by the phylum, Nematoda, popularly called the "roundworms." (Again, many of these are parasitic, but many are free-living.)

The roundworm invention is a fluid within the coelom which can slosh back and forth through the nooks and crannies of the organism. Food and oxygen can now be secreted into the fluid by those cells which absorbed an excess from the gut, and the fluid will carry it to all the cells it bathes for either immediate use or for storage. Similarly, wastes can be dumped into the fluid which can then carry it to the cells of the excretory system.

In short, the roundworms invented blood. The blood was an internal bit of the ocean which could bathe all the cells in an organism, however deeply buried. While a cell had an "ocean front" on the blood, it did not need to worry about the real ocean outside. It could rely on nourishment from the blood. That is why the roundworms could develop bulk and be round, whereas flatworms could only be flat.

The roundworms are also responsible for another advance. In both coelenterates and flatworms, the gut is a simple sac with only one opening. The indigestible residue of food taken in had to be ejected by the opening through which it had orginially entered. While ejection was taking place, further ingestion could not take place—and vice versa. They operate on the "batch" system.

The roundworms added a second opening to the gut, one in the rear. The roundworms were the first form of life to adopt the basic plan of a tube within a tube. Food particles enter at one end, are digested and absorbed as they travel along the gut, and the indigestible residue is ejected at the other end. Both ingestion and ejection can be continuous and, obviously, this moving-belt, continuous-process, assembly-line techinque of feeding represents another major improvement in the control of the environment.

Now from the roundworms, one can picture three different and important phyla branching off. Each one keeps everything possessed by the roundworms and adds a few novelties of its own.

In the first place, although the roundworms had the potentiality of bulk, thanks to the invention of blood, another obstacle remained in the way of complete realization of this potentiality. Roundworms are composed exclusively of soft tissue which must, somehow, withstand the disruptive effect of water currents. The larger an organism grows, the more vulnerable it is to this disruption unless it evolves some sort of stiffening.

This was invented by the phylum, Mollusca—or "mollusks," including clams, snails, oysters, et cetera. They developed a hard and rigid outer shell, or "exoskeleton," of calcium carbonate, which served several purposes. It stiffened the body and made more bulk possible. It served as a shield against enemies, and it served as an attachment point for muscles so that mollusk muscles could exert a far greater pull than could those of the flatworms or roundworms.

A second phylum tried a stiffening agent after another scheme. This was the phylum, Echinodermata—or "echinoderms," such as starfish, sea urchins, and so on—which developed a hardened shield under the skin, thus forming an internal skeleton or "endoskeleton." (Echinoderms seem to have retreated from the bilateral symmetry originated by the flatworms and to have returned to the radial symmetry of the coelenterates. This is actually a secondary modification. The larval echinoderms are bilaterally symmetrical and only take on radial symmetry as adults.)

In both phyla, the skeletons freed the organisms from some of the stresses of the environment to which the roundworms were subject. For this reason, both mollusks and echinoderms can be looked upon as advances over the roundworms.

However, the development of skeletons involved serious shortcomings, too. Mollusks and echinoderms are bulkier than flatworms and roundworms, to be sure, but the weight of their armor deprives them, by and large, of the free motion so painfully developed by animals. In place of wriggling worms, you have the relatively motionless starfish and oyster.

(Incidentally, general statements about phyla, or anything else, are not to be mistaken for universal statements. For instance, the most advanced of the mollusks are the octopi and squids which are anything but motionless. They have regained free motion, however, by abandoning the shell, except for vestigial remnants, and using other types of stiffening at strategic points.)

Again, a shell is a form of static defense. It brings about a kind of "Maginot" psychology. The animal retreats into a fortress and seems rarely capable thenceforward of elaborating refinements in its body plan that would involve an attack upon the environment. And it is always through an attack that the great victories in evolution are won.

Then, too, the shell is a wall that inhibits the creature from knowledge of the world. It is less bombarded by stimuli, thanks to its protective insensitive shell, hence is less apt to develop fast and accurate responses.

And yet the stiffening shell has advantages that more than compensate for all these disadvantages and it remains only to adapt it better; to keep its advantages while minimizing its disadvantages. I'll return to this.

But first, there remains the third development from the roundworms; one that does not involve a skeleton of any sort and is, perhaps, the most important of the three. This new advance shows up in the phylum, Annelida—or "annelids," of which the common earthworm, is the best-known example. The advance is "segmentation."

An annelid is composed of a series of segments. Each segment may be looked upon as an incomplete organism in itself. Each has its own nerves branching off the main nerve stem, its own blood vessels, its own tubules for carrying off waste, its own muscles, and so on. In designing a body plan which is a repetition of similar units, the forces of evolution are once again displaying the assembly-line philosophy, with a consequent improvement in efficiency. The annelid body scheme is better organized, flexible and efficient than is that of any nonsegmented creature.

Perhaps because of this, annelids could make further advances. For instance, they improved the blood system by inventing the just-mentioned blood vessels. Blood no longer sloshed back and forth in the coelomic cavity. Now it was confined to vessels through which it might circulate in organized fashion—more efficient. The annelids also invented hemoglobin, a protein which could carry oxygen with far greater efficiency than could a simple water fluid. (Yes, sir, the earthworm is entitled to considerable respect.)

Yet for all this, the annelids lack a skeleton. They remain soft and relatively defenseless and are limited in potential bulk. (Even the famous six-foot earthworms of Australia remain long and thin.) Their control of the environment is sadly limited.

So the next step is to develop phyla that combine the efficiency of segmentation with the security and bulk-and-strength potentialities of skeletal development. This was done no less than twice.

From the annelids—probably—there developed the phylum of Arthropoda—the "arthropods," including lobsters, spiders, centipedes, and insects. These retained the segmentation of the annelids but added to it the notion of the exoskeleton, originated by the mollusks.

The arthropod exoskeleton was, however, a great improvement over the mollusk exoskeleton. The former was not an inorganic compound, hard, brittle, inflexible. Instead, it was an organic polymer, called "chitin," which is lighter, tougher, and more flexible than the calcium carbonate shell of the mollusks.

Moreover, the arthropod exoskeleton was more than a shapeless barrier against the outside world. It was segmented, fitting the contours of the body closely, and therefore limiting bodily movements far less. In almost every way, chitin offered the advantages of the mollusk shell without the disadvantages. Add to this the efficiency of segmentation, and the arthropod body scheme obviously offers an advance over both the annelids and the mollusks.

A second phylum arose, probably from the echinoderms, at a time after they had invented the endoskeleton but before they had developed the adult regression to radial symmetry. The new phylum is Chordata—the "chordates" to which we belong.

The chordates retained the endoskeleton, which they gradually improved. They converted the primitive shell-like affair of the echinoderms into a system of internal girders which were comparatively light, impressively strong, and enormously efficient. They combined this with the introduction of segmentation.

You may be surprised to find out that you, as a chordate, are segmented, but you are just the same. The segmentation is not as outwardly visible among the chordates as among the other two segmented phyla. Among the annelids, for instance, it is clearly visible in the earthworm, and among the arthropods it is clearly visible in the centipede. However, though not clearly visible among the chordates, it is there.

Even in the human being who seems, outwardly, all one piece, a minute examination of muscles, blood vessels, and nerve fibers, clearly shows the existence of segmentation. The excretory and reproductive system in the chordate embryo—even the human one—shows clear segmentation, though this is obscured by secondary changes in the adult.

And you can see it yourself by feeling your backbone. Each vertebra represents one segment. This is most dramatic in the chest where each segment possesses not merely a vertebra but also a pair of ribs. (Or look at the skeleton of a large snake if you ever get the chance, and see if that example of chordate skeletal construction does not remind you of a centipede.)

That ends the march of the phyla, which is summarized in Figure 2, where once again the arrows do not necessarily represent lines of descent but do represent the direction of increased control of the environment, hence increased "advancement." There is no question in anyone's mind but that the arthropods and the chordates between them are the most advanced and important of the phyla. Again, if you wish, they "rule the world."

Their rule, in fact, may be permanent, for I wonder if any new phyla will ever be formed. Certainly, no new ones have been formed in a long, long time.

Life may have started three billion years ago and probably spent more than half its existence in the unicellular form. With the major discovery (or "breakthrough"—to use the currently fashionable phrase) of multicellularity, there may well have been an explosive exploration of the various versions of multicellularity. By the time the earliest fossils appear, all twenty-one phyla were probably already established.

Even the chordates and arthropods, the last to be established, were probably in existence in primitive form, at least 600,000,000 years ago—and no new phyla have been formed since.

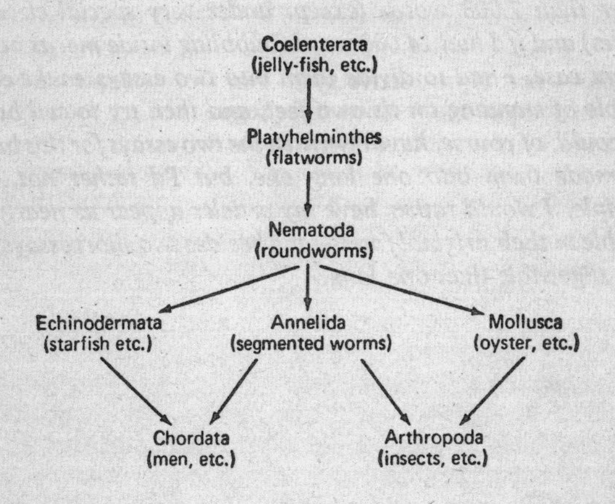

Figure 2

Does this mean that life holds no chance for improvement? Not at all.

For one thing there is much room for further advance and refinement within the arthropod and chordate phyla. For another, if the march of the phyla has ended, it may be that the potentialities of multicellularity have been exhausted.

Life may be readying for the step beyond the phyla, and that is what I want to talk about in the next article.

Clearly, this essay and the previous one were conceived as a unit and could have been written as a unit.

However, like all the essays I write, they are aimed first at some magazine, and these will only permit certain lengths, usually short ones, since they must contain material other than my own deathless prose.

Astounding *could not accomodate science essays that were longer than 7,000 words (except under very special circumstances) and if I had 14,000 words bubbling inside me, as in the present case, I had to divide them into two essays, make each capable of standing on its own feet, and then try to sell both.*

I could, of course, have rewritten the two essays for this book and made them into one long one, but I'd rather not. On principle, I would rather have my articles appear as nearly as possible in their original form—and besides two short essays are more digestible than one long.

3

BEYOND THE PHYLA

In the previous chapter—"March of the Phyla"—I concluded that there were two broad divisions—"phyla"—of living creatures that were more advanced than any others, in the sense that they had the greatest control over their environment. These two were the phylum, Arthropoda—or "arthropods," including lobsters, spiders, centipedes, insects and so on—and the phylum Chordata—or "chordates," including fish, snakes, birds, men, and so on.

I carefully did not try to make a decision as to which of the two was *the* most advanced. On the one hand, being men and, therefore, chordates, it seems natural to us that the chordates are the more advanced. On the other hand, it is undeniable that the mass of arthropod life in existence is far greater than the mass of chordate life.

Also, man may be ruler of the Earth, but he has certainly failed to control those insects that annoy him, despite heroic efforts. Annoying chordates, on the other hand, have generally gone down under the human onslaught; sometimes with embarrassing rapidity.

Perhaps that is the reason many of us have the unhappy feeling that when and if chordates—including even man—pass from the scene, insects—the most successful of the arthropods—will still be proceeding in quite business-as-usual fashion.

Despite any uncertainties we may have, however, if we restrict ourselves to the chordate individual and the arthropod individual, it is strictly no contest—with the chordate the obvious winner.

To see why, consider life on land.

Land life is a rather minor offshoot of life in general, since something like five-sixths of the total mass of living matter dwells in the oceans. However, control of the environment, which is the measure of the "advancement" of an organism is potentially possible to a much greater extent on land than in the sea. Consequently, land life has the odds very much in its favor in the competition for dominance. Why this is so is simply explained.

Life in the sea is surrounded by water, while life on land is surrounded by air. Water is about seventy times as viscous as air at ordinary temperatures and is that much more difficult to move through. There's the key point.

A creature capable of rapid motion is in better control of its environment and, therefore, more advanced—all other things being equal—than a creature not capable of rapid motion. But a sea creature designed for rapid motion must be streamlined, otherwise an impractical amount of energy is consumed in overcoming watery resistance. Examples of streamlining are to be seen at a glance in the sharks and fish.

Creatures on land, however, may be designed for rapid movement through the much less viscous air without being streamlined. Still, when the descendants of a line of non-streamlined land creatures return to the sea, streamlining sets in. You can see a little of it in otters and ducks, more of it in seals and penguins, and the near perfection of it in the porpoises and whales.

The disadvantage of streamlining is this: it discourages the existence of appendages which would break streamlining and destroy efficiency of motion. But it is by use of appendages that creatures can best handle the environment and bend it to their will. An opossum uses its tail to hold on to a branch, an elephant its trunk to manipulate large and small objects, a raccoon its paws, a monkey its hands, and so on.

In short, a streamlined creature cuts itself off from attack upon the environment. The whale is the most dramatic example of this. The whale is one of the two types of creatures with a brain that is larger than the human brain. The other type is the elephant, an undeniably intelligent animal.

The brain of the whale, unlike that of the elephant, is not only larger than a man's brain, it is also more intensely convoluted. There is a reasonable possibility therefore, that a whale might be—potentially, at least—more intelligent than a man. After all, porpoises and dolphins, small relatives of the whale, are undeniably intelligent, more so than most mammals, and a porpoise to a sperm whale may be as a monkey to a man.

But suppose a whale were potentially more intelligent than a man, how could it show that intelligence? It has a tail and two flippers that are perfectly adapted for powerful swimming and for nothing else. It has no appendages with which to manipulate the outside world and, thanks to the necessity for streamlining, can have none. What intelligence the whale has must remain strictly potential; a prisoner of the viscosity of water.

Or consider the giant squid, a member of the phylum, Mollusca. Certainly in all the world there is no more highly advanced creature that is neither arthropod nor chordate. In some ways, in fact, it goes all arthropods and chordates one better. It has large eyes, for instance, larger than any others in the world, similar to and possibly potentially better than the eyes independently invented by the chordates.

The squid has ten appendages, in the form of tentacles, which can writhe like so many snakes, each exquisitely sensitive and equipped with vacuum disks for tight grips. Yet these do not affect the streamlining, for when the squid chooses to put on speed, its streamlined mantle cleaves the water while the tentacles trail behind without interfering. In fact, since the squid moves rapidly by jet propulsion, it doesn't even need the fins or flippers that, in sharks, fish, or whales, unavoidably break the perfect pattern of streamline.

But the viscosity of water is nevertheless victorious, even over the superflexible arrangement of the squid. Those tentacles must move through water when they manipulate their environment and can only do so in slow motion. (Try swinging a bat under water and you'll see what I mean.)

To summarize then, the appendage is rare in the sea, and the quickly moving appendage nonexistent. The quickly moving appendage is, however, common among land creatures, and it is that which makes land species, not sea species, lords of the Earth.

However, there are also disadvantages involved in living on land. One arises in connection with gravity. In the sea, thanks to the buoyancy of water, gravity is virtually nonexistent. It is almost as easy for a fish to swim upward as downward.

On land, however, the pull of gravity is just about undiluted by the tiny air-buoyancy effect for any creature at the multicellular level. All living creatures invading the land must cope with it one way or another.

Until arthropod and chordate came along, all types of animal life that invaded the land were defeated by gravity. They coped by surrendering and moved on land only by slowly crawling, with the body in contact with the surface at all or almost all points. Watch an earthworm.

The development of shells by mollusks, which in the sea represented an advance, was, on land, actually a handicap. The land snail must not only fight the effect of gravity on its own body; it must carry the weight of a shell upon its back.

Now a crawling creature which needs all its energy to move in slow, ungainly fashion can scarcely develop fast-moving appendages. Therefore, the prime advantage of land life is lost for them. Below the level of arthropod and chordate, then, the most advanced forms of life are in the sea.

To develop fast-moving appendages, a land creature needs supporting legs to lift the main portion of the body clear of the ground in defiance of gravity. But legs of soft tissues alone will never support a body of even moderate bulk. Legs need stiffening. Both arthropods and chordates include types of creatures with stiffened legs; and to decide which of the two types is more advanced, let's find out which type uses the more efficient type of stiffening.

In the case of the arthropod, the stiffening is on the outside of the leg in the form of chitin. In the case of the chordate, it is on the inside of the leg in the form of bone. In general an exoskeleton—one on the outside—is the better for purposes of defense. An endoskeleton—one on the inside—is the better for structural strength. (Thus, a knight wears his armor on the outside and a skyscraper wears it steel girders on the inside.)

Furthermore, an exoskeleton limits growth. If the soft tissues within grow, the hard unyielding exoskeleton must be discarded, or growth must stop. In the arthropods, the exoskeleton is periodically discarded and replaced by a new and larger one. A great deal of vital energy goes into the perpetual manufacture of exoskeletons. What's more, during the interval of molting, the organism is fairly defenseless.

An endoskeleton does not limit growth. The bones within may freely be extended by accretion and the soft tissue about it yields, and, in fact, matches the growth easily.

The chordate individual can, therefore, grow larger than the arthropod individual and be stronger. Chordate muscles, slung on internal beams rather than on an external shell are more efficient. By all odds, the larger, stronger, faster chordate has better control of its environment and is, therefore, more advanced than the arthropod.

(Don't be fooled by stories to the effect that grasshoppers can jump so many times their own length and that ants can lift so many times their own weight and that if either were the size of a man it could do wonders. Actually, if either were the size of a man and could remain alive, it is quite certain that a grasshopper could not drive his bulk in a jump that was as long as a man's, nor could an ant lift as much as a man could.)

To be sure, not all chordates are equally advanced. The phylum, Chordata, is divided into nine classes, and of these, the first three include degenerate descendants of very primitive chordates. These now-living descendants rather resemble worms and mollusks outwardly, and only a zoologist would find reason for placing them in the same phylum with ourselves.

Nevertheless, these primitive creatures—or their more respectable ancestors—first stiffened their bodies with an inner rod of cartilage—a substance resembling chitin in terms of flexibility and toughness, though quite different chemically.

In addition, the first chordates apparently invented segmentation as well as hemoglobin, both of which were earlier and independently invented by the Annelida—the phylum to which the earthworm belongs. They also made entirely novel advances by developing a liver, in which many of the chemical tasks of the body were efficiently concentrated, and gill arches, which made respiration more efficient.

But this, obviously, is not particularly designed to make gravity-conquering land life possible.

The next class of chordates, Cyclostomata—of which the lamprey is the most familiar example—made a step in that direction by extending the one stiffening rod of cartilage into a complete skeleton, thus much strengthening the body and making it far less wormlike. In addition, they invented eyes—independent of the mollusk invention. The circulatory system also underwent improvements: a two-chambered heart was developed to drive the blood through the blood vessels and blood cells were developed in which hemoglobin might be kept. Both advances made the transport of food, oxygen, and waste matter more efficient.

Next comes the class of Pisces. This is divided into several subclasses of which the most primitive, Elasmorbranchii—sharks, et cetera—invented some of our most useful conveniences: jaws and teeth and two pairs of limbs.

The skeletons of both the lampreys and the sharks, while complete, are only of cartilage. This is sufficient stiffening for life in the water—the sharks are quite successful at it, too—but is not strong enough to support a moderately bulky creature against the gravitational force it would meet on land.

But another Piscian subclass, Teleostei, made use of a method whereby the skeleton was reinforced by inorganic salts such as calcium phosphate. Cartilage was thus converted to bone and the Teleostei are the "bony fish."

Further changes are necessary for land life. An organism must be able to utilize the oxygen of the atmosphere directly. In this direction, the teleostian had invented an air bladder by which it could increase or decrease its buoyancy at will, thus helping in vertical swimming. In some members of another Piscian subclass, Crossopterygii—the "lungfish"—the air bladder became a lung.

The crossopterygians are an example of the fact that it is often a loser in the game of life that makes a major advance. The crossopterygians were, for one reason or another, less successful at coping with their environment than the teleostians were. Most of the crossopterygian species are now extinct. Some descendants still exist by learning how to make do with environments so undesirable that the teleostians had no reason to follow them there, successful as the latter were in the lush pastures of the open sea. The crossopterygians retreated to stagnant water, to the oceanic abyss—and to the land. We are descended from the third group.

The next class of Chordata is Amphibia—of which frogs and toads are the best-known modern representatives. They made the transition. Amphibian lungs, working full-time in adult life, were given a circulatory system of their own, which made a three-chambered heart a necessity. In addition, the amphibians invented the ear. (In general, air being more transparent than water, sense impressions reach out farther into the environment on land than in the sea. Land creatures could more profitably sharpen their senses than sea creatures could. Sharper senses imply an increase in control of the environment and this, too, helps make land life more advanced than sea life.)

So it came about that amphibians were the first chordates to invade land, raise their bodies up on legs, and walk. They walked slowly and clumsily, to be sure, but they walked.

Toward the end of the Paleozoic era, the chordate amphibians and the arthropod scorpions and insects competed on land, and for the first time, a chordate victory began to show signs of inevitability.

But amphibians were still tied to the sea, or at least to a watery environment of some sort, during the period of birth and early development. It was the next class, Reptilia—the reptiles, that made the crucial invention: an egg that could be hatched on land.

Such an egg had first to be enclosed by a membrane that was porous to gases—so that the developing embryo could breath—but that could retain water so that the embryo would not dry up. In order for such an egg to be fertilized, fertilization must take place before the shell was developed and, hence, sperm had to be released within the female and not merely over the already laid eggs.

Again, the egg had to be large enough to contain the food and water needed by the embryo during the entire period of development. This meant the embryo must develop special membranes with which it might handle the food contents of the egg.

The reptiles developed all this and became truly a land animal. Some of them also put the final touches on the circulatory system by developing the fourth—and last—

chamber of the heart so that two complete and coordinating blood pumps existed.

The reptiles reached their heights in the Mesozoic era when giant dinosaurs shook the earth.

But the conquest of gravity meant that only one of the disadvantages of land life had been conquered. There was another one: temperature variation.

The temperature of the sea is virtually constant. Through almost all of its volume this constant temperature is fairly close to the freezing point. In a thin surface skin in the tropics, it is higher but, in that restricted area, still moderately constant.

Once a creature adapts itself to the temperature of its region of the sea, it needs no further adaptation to cope with changes.

On land, however, temperature varies widely. Land creatures can try to avoid that by living under rocks, in crevices, in burrows or in caves, by moving south in the winter and north in the summer, by hibernating through cold weather or by estivating through warm weather. These are all withdrawals, however, and avoidance mechanisms.

Success lies always in the direction of the offense. It was necessary to invent a device which would insure constant temperature within the body whatever—within reason—the temperature might be outside the body.

Two different groups of early reptiles made the necessary discovery independently, even before the great age of Reptiles had begun. One group developed into the class of Mammalia—mammals, like us—and the other, somewhat later, into the class of Aves—birds. Both had internal air conditioning, a way of storing heat in such controlled fashion that body temperature was kept within narrow limits.

In both cases, the body temperature was maintained considerably higher than the usual temperature of the environment. There was reason for this, since chemical reactions—and therefore the bodily movements that result—speed up with higher temperature. The highest temperature at which there is not too much damage to delicate protein molecules therefore implies best control of the environment and highest advancement.

But to maintain a high body temperature meant cutting down the rate of loss of heat to the atmosphere. This was done by keeping a layer of relatively motionless air next the body—still air being one of the best insulators.

The birds do this by trapping air among a set of modified scales called feathers; mammals by trapping it among a set of modified scales called hairs. (The feathers are the more efficient of the two, by the way.)

The birds took to the air, rediscovering three-dimensional travel which the amphibians had lost on leaving the sea. In doing so, however, birds found that the aerodynamic facts of life limited their size drastically, thus also limiting their potentialities for advance. Flight also involved the thorough commitment of one pair of limbs to the formation of wings—beautiful for their job but for nothing else.

So the future lay with the mammals which retained all four limbs relatively uncommitted and retained, furthermore, the possibility of large size.

The advantages of mammals over reptiles were, eventually, decisive. By possession of a constant internal temperature, they could be in full activity during the night and during the cold seasons, when reptiles were sluggish and at a relative disadvantage.

The possession of hair, moreover, meant the exposure of a soft skin to the environment and this is important.

The early chordates made a number of attempts to add, over and above the internal stiffening of bone, an external shield of some sort. The temptation to seek out protection is apparently almost irresistible. The earliest fish were armor-plated as were the early amphibians and reptiles.

The cost was too high in every case. The armored creatures only succeeded in making mollusks of themselves. Armor decreased the all-important mobility; it substituted passive defense for offense, which was unhealthy; and it put up a barrier between the outer world and the inner organism. The armored creatures invariably fell before the onslaughts of the unarmored ones. The last survivals today are the turtles, which are the most primitive and, on the whole, the least successful of the reptiles in existence today.

By converting scales into hair, the mammals became that much more sensitive to their environment, that much more able to respond to it, and, by responding, controlling it. Some early mammals made one last attempt to develop an external armor and went under. Their remaining descendants are the armadillos.

Temperature control did one more thing for mammals—and

birds, too. It made necessary the invention of extended child care. Or, if you care to be more dramatic, warm blood invented mother love.

Temperature control is more easily maintained in a large organism than in a small one. (All parts of the mass of an organism produce heat, but heat is lost only at the surface. A smaller creature has more surface per unit volume, hence loses heat at a greater rate.) This means that the most critical time in the life of a mammal, as far as heat control is concerned, is when he is smallest, when he is young or, most of all, when he is an embryo.

A sea creature can leave its eggs where laid and go away. The constant temperature of the sea will take care of them. A land creature without temperature control can take rudimentary precautions. A turtle, for instance, might bury its eggs in the warm sand and leave matters to the somewhat uncertain sun.

A land creature with temperature control—a bird, for instance—can't fool around. Its eggs require not only warmth but a certain constant temperature. There is not enough living tissue within the egg to supply that temperature, so it must be supplied from without; specifically, from the mother's body.

Under conditions of temperature control, then, survival of the species requires the development of instincts that will keep birds building nests, incubating eggs to the hatching stage, and feeding the young—all at considerable inconvenience to themselves.

The result, however, is a sharp decrease in infant mortality among birds as compared with reptiles. Inasmuch as the young bird is relieved of certain environmental stresses to which young reptiles are subjected, this represents an evolutionary advance in birds over reptiles.

The mammals go even further—in stages. The class, Mammalia, is divided into three subclasses. The first is Prototheria, which includes the duckbill platypus. These still show many reptilian characteristics and are imperfectly warm-blooded, but they have hair, which no true reptile has.

The Prototheria lay eggs, as reptiles do, but the embryo has proceeded in its development quite a way by the time the egg is laid, so that the incubation period, with all its special dangers, is cut down.

Furthermore, the Prototheria are the first to invent a special food supply for the infant, adjusted perfectly for its nutritional needs. This is milk, formed in the body of the mother and fed to the child via special "mammary glands"—whence the word "Mammalia."

The next subclass of Mammalia is Metatheria which includes marsupials such as opossums and kangaroos. Here another step is taken. The laying of the egg is so long delayed that it hatches first. It is an embryo at an early stage in its development that actually emerges. These embryos have just enough strength to make their way to the mammary glands of the mother, which are usually enclosed in a special pouch. In this pouch, the young complete their development.

The third and last subclass of Mammalia is Eutheria, or what we call the "placental mammal." Here the young develop to a much greater extent within the body of the mother. A special organ, the placenta, is developed, through which the developing embryo can absorb food from the mother's circulatory system and into which it can discharge wastes. This makes a longer gestation period possible; periods long enough in some cases so that the young are born almost capable of caring for themselves.

The development of mammary glands in the platypus reduces the environmental stress on the young to a level even below that among the birds. The pouch among the marsupials lowers it further, and the placenta among the placentals lowers it still further.

The comparison shows itself plainly in the fact that where the three subclasses of mammals competed directly, the placentals won out. Except for a few species of opossums in the Americas—where they persist by sheer powers of fertility—the only egg-laying mammals and marsupials that remain are those in Australia. Australia separated from other land areas before the placentals developed. Elsewhere, where placentals entered, the others went out.

The placental mammals are thus the current rulers of Earth.

Again, not all placental mammals are equally advanced. One thing that marks them off from one another is the development of the brain. Even the simpler mammals surpass in brain power the rest of organized life, but some mammals are brainier than others.

The good brain development of the mammals is probably the consequence of life on dry land, on soft skin, and the improvement of the sense organs generally. Mammals were, in consequence, buried under a large mass of sense impressions and there was consequently survival value in the further development of an accounting system—so to speak—to sort out those impressions and devise responses.

But one thing further is needed. There is still the question of appendages, which is the greatest gift of life on land. But to be maximally useful, an appendage must be useful in a variety of ways. There is always the danger of overspecialization.

Thus, I have mentioned the wing of the bird. It is a fast-moving appendage, but it can do only one thing. Similarly, the marvelously organized legs of horses, deer, and antelope are excellent devices for outracing the enemy, but they are no longer useful for anything else.

On the other hand, the raccoons and bears walk flat-footedly in primitive fashion—as we ourselves do—and their paws can be used for a variety of tasks. The members of the dog family, and also some of the rodents, retain the ability to use their paws as exploring devices. The elephant has developed a trunk that is the nearest thing any land creature owns to the tentacle of a squid.

The use of any such appendages increases the number of sense impressions the animal must handle. Again the brain enlarges and its ability intensifies. (The whale is an exception; it has a large complicated brain with no generalized appendages. Perhaps its brain is a legacy from an intelligent land-living ancestor—nothing is known of the ancestors of the whales, after all. Or perhaps it is only a response to the need for coordinating fifty to one hundred fifty tons of living matter.)

Obviously, the variously useful appendage reaches a climax in the order of Primates—the monkeys, apes, and us—in which at least two, and sometimes all four, of the limbs end in hands in which the individual fingers are capable of more or less independent motions. In the more advanced members of the primates, one of the fingers, the thumb, is well developed and faces the other four, converting the hand into a possible pincer.

The Primates are, not surprisingly, the brainiest of the mammals, and man, with the best-developed hand, is, not surprisingly, the brainiest of the Primates.

By using his brains, man was able to extend the two most fundamental inventions of land life generally. He learned to control fire and thus extended the notion of warm-bloodedness. Other mammals and birds might control the internal temperature, man controlled the external temperature as well. Man also developed the systematic use of tools, which equipped him with artificial fast-moving appendages, each of which might be thoroughly specialized. He gained all the advantages of specialization without abandoning any of the advantages of nonspecialization.

And so man is the lord of the Universe and—

Where do we go from here?

It is possible to imagine a bigger and better man, a "superman," but that need not be the answer. Bigger and better dinosaurs just ended in extinction. Bulk alone is not everything. Neither is brain power alone.

Actually, multicellularity may be played out. It may be that the multicellular organism has reached its limit. There has been no new phylum of organisms established in perhaps 600,000,000 years. Within the phylum of Chordata—the last-established—there has been no new class established in at least 250,000,000 years. Within the class of Mammalia—the most advanced of the chordates—nothing better than the placental mammal has been established in 100,000,000 years.

The great experiments may be over. What we are now facing is merely a refinement and a re-refinement of existing experiments.

But all this has happened once before.

A billion years ago, one-celled life had reached its peak. After many victories, such as the discovery of food storage and of photosynthesis, cells reached their limits. Evolution came to a dead end, or would have but for an entirely new breakthrough. Cells developed into cell colonies and then into multicellular organisms.

Now multicellularity has reached its dead end, too. Is there room for a new breakthrough? Can there be, once again, a new combination to a higher order of creature, a multiorganismic being. Such a combination must be more than merely physical, since physical combinations would just make a larger multicellular organism. (In fact, the physical combination of organisms was tried, after a fashion, with the invention of segmentation. It was an advance but not nearly as fundamental a one as multicellularity.)

Fortunately, we have examples of nonphysical combinations of multicellular organisms.

Many varieties of creatures herd together in groups that act with a certain primitive coordination. They move together, feed together. If one is frightened, all flee. They may even combine for protection against a common enemy—though generally they merely run and devil take the hindmost. Or they may combine to hunt prey and then, often, quarrel over the spoils.

Such herds, or packs, or schools are the equivalent of cell colonies on the cellular level. Although it may be convenient for groups to keep together, it is not vital. Each individual in the herd can, if necessary, survive on its own.

We must look for something more than that.

In the previous essay, I used one main criterion to distinguish between a multicellular organism and a mere cell colony. In a multicellular organism, individual cells become so specialized that they can no longer live independently and the component cells are subordinated to the group to the point where only group-consciousness exists.

No group of organisms display these characteristics to the full, but there are signs of beginnings. The clearest cases are among the phylum, Arthropoda, and in its most advanced and most recently established class—Insecta, the insects.

The three main groups of "social insects" are the bees, the ants—both belonging to the order, Hymenoptera—and the termites—belonging to the order, Isoptera. All three display specializations among constituent organisms just as multicellular organisms display specializations among constituent cells. In the case of termites, the specializations go so far as to make life impossible for certain individuals outside the society—one of the hallmarks of a true multiorganismic creature. The termite queen cannot live without her attendants. Termite soldiers have mandibles so large they cannot feed themselves. They must be fed by workers.

Furthermore, such societies are more advanced than any individual organism, not only of their own type but of any type. A society of even primitive individuals can beat even a very advanced individual who happens to be on his own. When the army ants go marching, there is only one way the big-game hunter—elephant gun and all—can save himself. He has to get out of the way, and fast.

There is a classic story called "Leiningen and the Ants," which tells of a plantation owner who found his land in the way of a marching column of army ants and decided to stand his ground and fight. Leiningen was a most superior individual, brave, resourceful, intelligent and he fought like a demon. He managed just barely to get out of the fight with his life.

You might think the odds were terrific—millions of ants against one human—but you'd be wrong. The odds were exactly even numerically: *one* man versus *one* ant society.

To be sure, lots of individual ants were killed but that didn't affect the ant society. Leiningen lost skin and blood, trillions of his individual cells, but he recovered and did not feel the loss.

Outside the class of insects and the phylum of arthropods, there is only one example of a society that begins to be more than an organism colony. That is, of course, the human society. It includes specialized individuals—not physically specialized, to be sure, but mentally specialized. Some of them are so specialized they cannot live outside the society—and there is the hallmark again.

I, for instance, am city-bred and have lived—with moderate success—as part of a complex society all my life. I eat only too well, alas, but I cannot raise food; I have no experience in gathering food, I cannot even cook. I drive a car, but do not even know how to lift the hood. I own a house, but cannot repair any part of it. I watch television and use a number of appliances, including an electric typewriter, but am helpless in the face of electrical wiring.

Without the continuing and intensive help of other members of the human society, I would not survive long. Alone on Robinson Crusoe's island, I could only hope for a quick death in preference to a slow one. I think there are millions like me.

But now what is it that holds a society together—a true society where the component individual is willing to die for the good of the society? In the case of the insects, it is something we call "instinct," a compulsive, robotlike behavior that deprives the individual insect of choice of action. The individual insect is not only *willing* to die for the group, it *cannot* do anything else.

But what holds together a human society? Certainly not instinct. The nearest thing we have to an instinct in the matter is one which says, "To hell with the others. Cut and run." Often, this instinct is obeyed. The surprising thing is that often it is not obeyed.

I said earlier that intelligence was not enough in itself. Obviously, if it is joined to other qualities that are disadvantageous, extinction will follow. An intelligent animal that is too limited in the climate it can tolerate, or the food it can eat, or the parasites it can resist, is not going to succeed. The elephant and the great apes are examples of intelligent failures.

But when the first hominid rose to his hind legs, what made him a success when the gorilla was and is a failure?

I say that for hundreds of thousands of years the early hominids were on the borderline of failure. It was the crucial breakthrough of the formation of a tribal society that really set them on the road to mastery. Not merely packs mind you, after the fashion of baboons, but a true society in which the whole was something more than the sum of the parts.

What made this possible, it seems to me, was the development of a means of communication that was complex enough and flexible enough to express abstract ideas—to be something more than a mere squeal of fright or a simple warning cry.

By means of such communication—peculiar, as far as we know, to *Homo sapiens*—the amassed learning of one generation could be passed on to another. A young man absorbed in his youth what had taken an old man all his life to learn, and then the young man went on to learn more on his own. A new and larger body of knowledge was passed on to the generation after.

But with learning from the old came a reverence for the old; a new feeling that only human beings could have—tradition.

"This is the way things are done; this is the way things have always been done; this is the way our ancestors said it should be done; and because their spirits watch us and must not be angered, this is the way things *must* and *will* be done."

There is no use belaboring the point. We all know the power of tradition. It will hold a society together almost as firmly as instinct will. Call it "duty" or "patriotism" or "altruism" and any one of us can bring himself or herself to the point where he or she will give up individual life for the good of the group—which might be a small one called the family, a larger one called the nation, or a still larger one called humanity.

And if it was oral communication that launched the tribe and the first cultures, it was written communication that launched the cities and the first civilizations into full flower.

But are the city and the anthill the final expression of the multiorganismic being? It seems to me most certainly not. Both are only at the beginnings of society potential.

The insect societies have succeeded, much more than has the human society, in physically specializing their members and in generalizing consciousness from the individual to the society. However, their method of doing this has cost them flexibility. Each individual insect in the society may make only the most limited responses to given stimuli.

The human society has specialized far less and has retained far more of individuality, but it has compensated for that by retaining a most successful flexibility.

The next step, it seems to me, would be the combination of the two—a society which combines an insectlike consciousness of the whole with a humanlike flexibility.

What type of organism, then, will attain this next major step in evolution?

To answer the question, let's look at the overall record of evolution so far. All through evolutionary history, it seems, once a particular type of organism has made a major advance, it is a subtype of that type and then a sub-subtype of that subtype that makes the next major advances. There is no coming from behind in evolution.

In other words, once the chordates are evolved and by dint of internal skeletons prove to be clearly more in control of the environment than are the mollusks, the die is cast. Further evolution merely increases the lead of chordates generally over mollusks generally. In the same way, land chordates increased their lead over sea chordates, mammals increased their lead over reptiles and humans over nonhumans. No group, having once relinquished the lead, ever gave rise to descendants that regained the lead.

Thus, at the phylum level, Chordata and Arthropoda are clearly in first and second place, respectively, from the moment of first clear-cut development half a billion years ago or so, and have never relinquished those positions. They are less in danger of relinquishing the lead now, in fact, than they have ever been before; a lead that is so secure that no new phyla have even been tried ever since the rise of the chordates.

Both phyla are divided into classes. Within Chordata, Mammalia lead all other classes. Within Arthropoda, Insecta lead all other classes. The mammals and insects have been increasing their lead ever since their first clear-cut development and are in less danger of losing it than ever before.

This process continues, as shown in Figure 3, where the

arrows do *not* indicate lines of descent but only the direction of increasing control of the environment. An underlining of a group of organisms symbolizes "dead end."

On the past record, then, it would seem that the next step would have to be taken by subdivisions of the "winners" of the last step; subdivisions, in other words, which are descendants either of the social insects or of man.

Now it seems to me that insects must be ruled out. In the first place, the insect societies are clearly in second place to human society as far as control of environment is concerned and there is no coming from behind in evolution. (Remember, I don't say that insects may not out-survive man despite this.) Secondly, they are already too specialized and too inflexible to reverse their ground and gain the necessary flexibility for a higher multiorganismic society. In evolution, specialization is invariably a one-way street and moves only in the direction of more specialization.

The only possible ancestor of the multiorganismic society, then, is man who is, physically, a relatively unspecialized animal except for his brain; and mentally, thanks to his relatively poor supply of instincts, is equally unspecialized.

The possibility that man will be the ancestor of the multiorganismic society is strengthened by the fact that he represents, for the first time in evolutionary history, an organism which is consciously aware of the competition of other organisms and will surely make a special effort to wipe out any new group which threatens his own overall superiority. Super-chimpanzees, unless overwhelmingly superior from the first, will almost certainly be erased as soon as they are recognized for what they are—barring some individuals retained for scientific observation.

So it might seem that, eventually, a family of human beings that are joined together on a new level into a multiorganismic society may be established—or, for all most of us know, has been. If they are not recognized for what they are too early, they will take over.

A more classic device for evolution is to suppose man to be divided into groups which are completely separated geographically, so that the gradual summing of mutations produces separate species no longer capable of interbreeding. One such new species may then develop the multiorganismic society and will then be to the remaining species as man is to the other mammals—or, perhaps, as man is to the amoeba.

Of course, on Earth there is no longer any chance for a complete geographic separation of any group of men and women over a period long enough to make that work, barring a devastating nuclear war that leaves only remnants of survivors and a completely disintegrated technology.

However, the time may be coming when colonies will be established on worlds other than Earth, on worlds outside the Solar system, perhaps. "Geographic" isolation may then be possible. Men venturing out into space may be like the crossopterygian fish venturing out onto land. They leave as experimenters and end as victors.

It is, of course, repugnant to a human individual at the present stage of his development to think of himself as a mere unit in a multiorganismic society, without will of his own and, whenever necessary, liable to be sacrificed, cold-bloodedly, to the overall good.

But is that the way it will be? It's extremely difficult to imagine what being a part of a multiorganismic society will be like, but suppose we consider the analogous situation of cells in a multicellular organism.

Component cells cannot live apart from the organism, but within the organism they maintain themselves as biochemical units. They produce their own enzymes, conduct their own reactions, have membranes that separate them from their fellows, grow and reproduce on their own in many cases.

In a multiorganismic society, the individual may well retain a good deal of mental and physical independence. He may think for himself and have his own individuality and *also* be part of the greater whole.

As for being sacrificed cold-bloodedly—not unless it were necessary. Skin cells die as a matter of course while the organism lives, but Americans die as a matter of course while the nation lives. Other cells may die on occasion for the good of the whole, but even in our imperfect society, so must policemen, firemen, soldiers, miners—

We do not allow our cells to be killed for no reason. Thanks to a sensation known as "pain" we take good care of our component cells and would not as much as endure a scratch or a pinprick if we could avoid it. A multiorganismic society would be as careful of its components and would undoubtedly feel something akin to pain at any harm to them.

And then there would be a positive gain. In passing from a cell to a group of cells, it becomes possible for the cell totality to appreciate abstract beauties such as those of a symphony or of a mathematical equation that the cells separately could never conceive of. There may be the cellular equivalent of these beauties in the waverings of a water current or in the engulfing of a tiny organic fragment, but who can argue that a man does not achieve a more exalted rapport with the universe than an amoeba can? Or what man can imagine that the individual cells of his body—which must share somehow in the complexity of his relations to the universe—would rather return to being just so many amoebas?

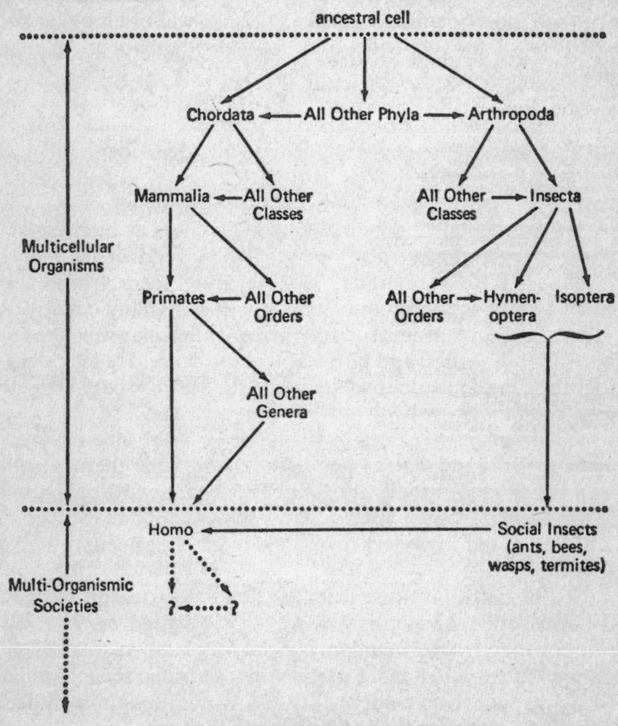

Figure 3

And, by analogy, who knows what unimaginable sensations, what new levels of knowledge, what infinite insights into the universe will become possible for a multiorganismic society? Surely there will be something then that will compare with a symphony as heard by a man, as that symphony compares with a wavering water current as felt by an amoeba.

It is impossible for me to write over a thousand articles on various aspects of science, as I have done, without duplication of information.

This is especially true when a magazine requests me to deal with a specific topic of their choice. It is then no use telling them that I have already dealt with that topic in another place at another time. They would quite rightly point out that only a vanishingly small percentage of their readership is likely to have seen the other article and that, in any case, they want the information tailored to their needs. So I oblige.

In the case of this essay, the information, or at least some of it, is present in an essay entitled "Our Evolving Atmosphere" which you will find in my essay collection Is Anyone There? *(Doubleday, 1967.)*

Is it fair, then, to include this essay in the present collection?

I think it is. This present essay is written from a different viewpoint and includes material I did not include in the earlier essay. The information may overlap, but the essay is different.

4

THE GIFT OF THE PLANTS

We tend to take our atmosphere for granted. We give little thought, if any, to the oxygen we breathe, the oxygen that is always there, ready for us to inhale some sixteen times a minute, and without which we could not live.

Most people these days understand, when they stop and think about it, that the oxygen in the air is the gift of the green plants. Plants form their tissues out of carbon dioxide and water and minerals, and, in doing so, discard some of the oxygen and give it off into the atmosphere.

The oxygen that is their gift has an importance over and

above its breathability, however. It may have made land life possible on Earth, so that this, too (and we ourselves), is the gift of the plants. To see how that might be, let us go back to the beginning of life on this planet, a beginning that seems to have taken place well over three billion years ago.

At that time, there was no oxygen in the atmosphere. Earth had a "reducing atmosphere" then, one which contained hydrogen—both alone and in combination with other elements. This is only natural, since the material out of which the Solar system formed was about 90 per cent hydrogen.

Hydrogen itself could not be retained in quantity because its molecules are so light, but that left the hydrogen combinations with oxygen, with carbon, and with nitrogen. These combinations formed molecules of water, methane, and ammonia respectively. There was a water ocean, with ammonia dissolved in it, and the air above was chiefly methane with vapors of ammonia and water and possibly some hydrogen.

We couldn't live in such an atmosphere, nor could any form of plant or animal life flourishing on Earth today. Yet, oddly enough, it was out of such an alien chemistry that life first originated on Earth—in very simple forms, to be sure.

This is not a matter of simple guessing. For the last twenty years, scientists have been working with sterile mixtures of those compounds thought to exist in the Earth's atmosphere and ocean billions of years ago. They have added energy in the form of ultraviolet light to mimic the energy of primeval sunlight. As a consequence, they found that the simple molecules of the early Earth combined to form more complicated ones that seem to be on the road to life as we know it.

In the laboratory, the process has been followed only through its very beginnings, but it is easy to imagine what would happen in a whole ocean of compounds over hundreds of millions of years ago. Molecules would grow more and more complicated, until finally some had grown complicated enough to begin to have some of the properties we associate with living matter.

Ultraviolet light is a double-edged sword, however. Its energy can force the beginning of a process whereby small molecules combine to form larger, more energy-rich ones. The very large molecules associated with life, however, are "rickety," and the energy of ultraviolet light can knock them apart again.

Fortunately, the water of the ocean absorbs ultraviolet light. In the very uppermost layers only middle-sized molecules could be formed but a little lower, where the worst of the ultraviolet light does not penetrate, the very complicated molecules of life might persist. We get a picture of very simple organisms resting some distance below the surface during the dangerous day, then rising at night to feed on the smaller compounds which can survive the ultraviolet light and which can serve as food.

Life could not form in the water soaking the soil of the islands and continents, however; nor could life already formed in the ocean migrate to land. On land, there would be no easy way of escaping the deadly ultraviolet light. Bits of land life could not burrow into the soil as easily as bits of sea life could sink in water. For that reason, the land remained sterile through the early history of life on Earth.

Only relatively small amounts of life could exist under these early conditions; only the amount that could be supported by the food molecules formed in the upper layer of the ocean by ultraviolet light.

As time went on, things got worse. Molecules of water vapor in the upper atmosphere were broken apart by the energy of ultraviolet light and this initiated chemical changes that changed ammonia to nitrogen, and methane to carbon dioxide. Eventually, the Earth developed a new atmosphere. Atmosphere I, made up of ammonia and methane, changed into Atmosphere II, made up of nitrogen and carbon dioxide.

The kind of compounds that would serve as food do not form as readily in a nitrogen and carbon dioxide atmosphere as they would in an ammonia and methane atmosphere. In other words, as Atmosphere I changed slowly into Atmosphere II, the amount of food in the uppermost layers of the ocean thinned out.

Life might not have survived at all but for the development of a compound called "chlorophyll" which is what makes plants green. This may have taken place quite early in life's history. Atom combinations resembling chlorophyll have been formed in the laboratory out of the primeval compound mixture and there is evidence of chlorophyll-containing organisms (blue-green algae) among the very first signs of life we can find.

The advantage of chlorophyll is that, using the energy of visible light (not ultraviolet light), food molecules can be formed directly out of carbon dioxide and water. The action of ultraviolet light is no longer needed.

As Atmosphere I changed into Atmosphere II, then, those bits of life that depended on ultraviolet light to form food gradually died out. On the other hand, those bits of life with chlorophyll (which we now call "green plants") multiplied as the carbon dioxide content of the atmosphere grew.

Finally, when Atmosphere II was fully established, green plants were the dominant form of life on the planet. That may have happened less than a billion years ago. And even then, the land remained sterile, for ultraviolet light still blazed down on the unprotected land surface.

As green plants multiplied steadily during the changeover from Atmosphere I to Atmosphere II, they produced oxygen at a steadily increasing rate. This oxygen did not remain as such, but combined with the components of Atmosphere I to change them into Atmosphere II. In other words, not only did green plants benefit from the changeover, they actually hastened it.

After the changeover was complete, green plants continued to produce oxygen, but now oxygen had nothing to combine with and so it accumulated in the air. As time went on and green plants continued to multiply, it did so at the expense of the carbon dioxide in the air. The carbon dioxide grew less and the oxygen grew more. The atmosphere changed once again, this time into Atmosphere III, the nitrogen and oxygen atmosphere we enjoy today.

The presence of free oxygen in the atmosphere was crucial to life for the following reason:

In a non-oxygen atmosphere such as the Earth had until (perhaps) less than a billion years ago, living organisms obtained their energy by breaking down the middle-sized molecules of food into smaller molecules. The amount of energy obtained in this way, however, is rather small. This means that life forms could not display energetic activity. There just wasn't enough energy in the food to permit it.

The simple plant life of the sea does not display energetic activity even today, but there must have been, from pretty nearly the beginning, other forms of life. There must have been forms of life that could not manufacture their own food since they lacked chlorophyll and therefore had to subsist, parasitically, by eating plants. These were the first animals.

Potentially, animals could make use of energy at a greater rate than plants could. A single animal could eat many plants and make use, lavishly, of the food energy the plants had only very slowly accumulated. But even so, without oxygen in the atmosphere, the total amount of energy an animal could develop was small. Until less than a billion years ago, then, animals were no more complex than plants and hardly any more active.

But then, as the oxygen content of the atmosphere slowly built up, chemical mechanisms were developed within cells that made it possible to combine food molecules with oxygen in the process of breaking them down. This made an enormous difference in energy development. Food molecules, when broken down by way of combination with oxygen, developed about twenty times as much energy as those same molecules would have done if they had been broken down without oxygen.

Those animals that developed the ability to make use of the oxygen being poured into the atmosphere by the green plants found themselves with an incredibly rich supply of energy that could be used for many "luxury" purposes that would have been impossible before.

It meant that the simple animal organisms of Atmosphere II could grow more complicated and could develop tissues and organs that did not contribute directly to the process of eating and energy-gathering. In particular, they could develop hard parts for protection.

It is the hard parts—shells, bones, teeth—that most readily turn into stony substances with time, and it is these which are left behind in the rocks as fossils. The earliest fossil-rich rocks are of the Cambrian era, about six hundred million years old, and it is only after that that the history of life can be described in some detail.

The thought that, because the first fossils show up six hundred million years ago, life *began* then is obviously wrong, for the first fossils are those of organisms that are already almost as complex as present-day organisms are. They obviously had many hundreds of millions of years of evolution behind them. Actually, the first fossils appeared only after at least four-fifths of the history of life had elapsed.

The fossil record starts suddenly, at this late date, because it was only then that Atmosphere III had developed sufficiently to allow animals to develop hard parts. Till then, there had never been enough energy to spare for such a purpose. It was only with oxygen in the air and with the supply of energy increased twentyfold that there was an almost explosive development of animals into complexity.

At the time the first complex animals with hard parts were developing, it may be that the oxygen content of the atmosphere was still far less than it is now. The oxygen content continued to rise, however, in the atmosphere itself, and in the ocean (through solution), where the life forms dwelt, until almost all the carbon dioxide was used up.

For a period, even after the development of Atmosphere III permitted the appearance of complex animals, life *still* remained confined to the sea. The first third of the fossil record is of sea animals only. It is only four hundred million years ago that life began to colonize the land surface of the planet. It is only in the last eighth of the story of life on Earth that the land has ceased to be sterile.

If the sterility of the land was due to the dangerous ultraviolet light in the Sun's radiation, what happened four hundred million years ago to put an end to the menace? It may be that what happened was another aspect of the gift of the plants—oxygen.

An oxygen molecule, as it occurs in the atmosphere, is made up of two oxygen atoms in combination and is written O_2 in consequence. In the upper atmosphere, the energy of sunlight can add a third oxygen atom to form O_3, which is called "ozone." This means that an ozone layer forms in the upper atmosphere about fifteen miles above the surface of the Earth.

Naturally, an ozone layer doesn't form unless there is an oxygen supply in the atmosphere.

As soon as Atmosphere II began to turn into Atmosphere III through the activity of green plants, and oxygen began pouring into the atmosphere, ozone began to form in the upper atmosphere. At first, the ozone that was formed must have been tiny in quantity, but as the oxygen content grew in the atmosphere below, so must the quantity of ozone in the atmosphere above.

The ozone layer never grew very dense; it is not at all dense even today. Ozone, however, has the capacity to absorb ultraviolet light very efficiently. Even a thin layer of ozone will stop ultraviolet light as though by a brick wall.

This means that as Atmosphere II turned into Atmosphere III, the ultraviolet light reaching the earth from the Sun slowly decreased. This had no ill effect on green plants which, through chlorophyll, depended on the energy of visible light for their food, and visible light can pass through ozone easily.

By about four hundred million years ago, there must have been enough oxygen in the atmosphere to provide an ozone layer high above earth that was dense enough to stop most of the ultraviolet light from the Sun. It then became possible for life to exist even when exposed to the by-now-no-longer deadly radiation of the Sun.

First, plant life colonized the land higher and higher above the tidal level. Then spiders, insects, snails, and other small forms of animal life followed, feeding on the plants. Then the first vertebrates crawled out on land—first the amphibians who still had to return to the water to lay their eggs, and then the reptiles which, for the first time, developed large eggs capable of being hatched on land.

Land life, we must realize, was capable of advances fundamentally different from those in the sea. In the sea, organisms are surrounded by water, which has a relatively high viscosity. To move quickly in water, sea organisms must be streamlined, which reduces the possibility of complex appendages.

On land, animals are surrounded by low-viscosity air, which means that they can develop very irregular shapes and still move quickly. In particular, they can develop strong and complex appendages, even a limb that could develop a complex, versatile, and flexible hand at its end. It is the hand and eye of primates that make it possible for them to observe the environment keenly and handle it delicately, and that, in turn, stimulated the growth of of the brain and the intelligence.

Further, exposure to a free-oxygen atmosphere, rather than to a water ocean, makes fire possible, and from fire all the rest of technology arises.

So it boils down to this:

Green plants created the oxygen atmosphere that made complex animals possible.

The oxygen atmosphere, in turn, created the ozone layer that made land life possible.

Land life made possible limbs and hands, and the oxygen made fire possible.

And here we are—complex, land-living, hand-and-eye technologists—*Homo sapiens*, the gift of the plant.

As I mentioned in earlier essay collections, notably in The Beginning and the End *(Doubleday, 1977),* TV Guide *has the pleasant habit of asking me to do backgrounders for them now and then, and on a variety of subjects.*

Sometimes they set guidelines that are hard to meet. When a TV special on the human brain was slated to appear, they sent me an outline of the script and directed me to write a thousand words on some aspect of the brain that was not covered in the script. Fine, except that every aspect of the brain was covered in the script—or seemed to be.

I thought about it and finally realized that the evolution of the brain was not discussed. That's understandable. Mentioning anything about evolution will surely rouse the anger of what are perhaps the most articulate set of obscurantists in the nation. The TV people may quail at this but I don't, so that's what I wrote about.

After the article appeared, I naturally got a bundle of letters denouncing me for daring to deviate from the literal words of the Bible. It's rather a shame. Now that the creationists are deprived of their chance of burning people at the stake, their best argument is gone.

5

THE BRAIN EXPLOSION

The brain is by far the most complexly organized piece of matter we know. It is enormously more complicated in structure than a star is, for instance, which is why astronomers know so much about stars and psychologists know so little about brains.

That is also the reason, perhaps, that it took so long for evolving life to do very much with a brain. Such complexity takes time to develop.

The first scraps of life appeared on Earth perhaps 3,500 million years ago. About 100 million years ago (when 97 per cent of life's history had already passed) the giant reptiles we call dinosaurs were supreme. In many ways, they were the most magnificent creatures the Earth has ever seen—large and powerful, some of them enormous predators and some as armored as tanks, some flying, some swimming, some running, some undoubtedly agile and energetic.

And yet they were tiny-brained. Billions of years of evolution, and yet their skulls held almost nothing. The stegosaur, for instance, an armored two-ton monster, had a walnut-sized brain weighing no more than 50 grams (2½ ounces).

But dinosaurs died off 70 million years ago (for reasons that aren't clear) and the mammals succeeded to world mastery. For tens of millions of years they had skulked in the shadow of the dinosaurs; small, furtive, and almost as unbrained.

But once the mammals had the world to themselves, they multiplied, evolved in many directions, and, quite suddenly, the brain began to expand.

The expansion was most marked in that group of animals called "primates" and reached its peak among the larger species of the group—the great apes.

The weight of the brain of the orangutan is about 340 grams (12 ounces), nearly seven times as large as that of the stegosaur, even though the orangutan is a far smaller animal. The brain of the chimpanzee is 380 grams (13½ ounces) and that of the gorilla, the largest of the primates, is 540 grams (19 ounces or just about $1^1/_5$ pounds).

But if the gorilla is the largest of the primates, it is not the brainiest—for the human being also belongs to the group. In fact, the extinct, half-human predecessors of mankind were already setting new records. *Homo habilis*, a manlike primate who lived about three million years ago, had a brain as large as that of a modern gorilla. *Homo erectus*, who lived about one million years ago had a brain weighing about 1,000 grams ($2^1/_5$ pounds).

We ourselves, *Homo sapiens*, came on the scene about half a million years ago, and we do better still. A human baby at birth already has a brain weighing 350 grams, about that of a full-grown orangutan. An adult male human being today has a brain with an average weight of 1,450 grams ($3^1/_5$ pounds).

Some individual brains even reach the 2,000-gram mark ($4^2/_5$ pounds).

In other words, our brain has tripled in size in the last three million years and that is an explosive change by evolutionary standards.

Why? No one really knows. Perhaps as long as all animals are small-brained, having a slightly larger brain doesn't make much difference as far as intelligence is concerned and other facts guide the evolutionary direction. Once past a crucial size, however, intelligence becomes great enough to exert a commanding influence and then even small additional increases can have important survival value. Selection for increased brain size then becomes strong and steady.

Of course, the human being doesn't really hold the record for sheer brain size. The largest elephant brain ever weighed was 8,000 grams (16½ pounds) while a sperm whale brain was weighed at 9,200 grams (20¼ pounds). That 20-pound brain is surely the largest that ever existed.

Yet size alone isn't the sole criterion of intelligence. If a large brain must handle an enormous body, so much of it is taken up with the task that little is left over for the imponderable of abstract thought.

For instance, a stegosaur's brain is only $^1/_{25,000}$ as heavy as its body. A brain can't handle 25,000 times its own weight and do more than just keep the body alive. Yet a sperm whale with a 9,200-gram brain, about 180 times the weight of a stegosaur's brain, isn't much better off. After all, a sperm whale is about forty times as heavy as a stegosaur and its brain is $^1/_{6,000}$ as heavy as its body. In the elephant, the ratio is $^1/_{1,200}$.

Compare this with the ratio in the human being—$^1/_{50}$. What it amounts to is that each pound of human brain has only $^1/_{150}$ as much body to worry about as a whale's brain does and only $^1/_{20}$ as much as an elephant's brain.

An adult woman's brain is on the average only 90 per cent as heavy as an adult man's brain. The woman's body is *less* than 90 per cent the weight of the man's body, on the average, so her brain/body ratio is a bit higher than a man's. Make of that what you will.

Actually, the human being doesn't hold the record for brain/body ratio either. The smaller monkeys do. The marmoset has a brain/body ratio of $^1/_{18}$. If a human being had *that* brain/body ratio, his brain would be half the size of an elephant's.

However, the total weight of a marmoset brain is only about 50 grams at most. It just isn't large enough to contain the number of brain cells required to make it capable of abstract thought.

The human being strikes a happy medium, then. Those few animals with brains absolutely larger than our own have such enormous bodies that the brain can't handle it all and have much left over for sheer intelligence. Those few animals with brains proportionately larger than ours are so small that the brain is too tiny to be intelligent.

So we stand alone. Almost. There are competitors.

There are dolphins and porpoises, small members of the whale family, that weigh no more than human beings and yet have brains slightly larger than those of human beings.

Does that make them of human intelligence? We can't say. Experimenters who have worked with dolphins have been unable to cross the species boundary and penetrate the workings of the dolphin mind.

But that's not surprising. We can't really understand our own brain. How, then, can we understand the dolphin's.

I have a tendency to view man as the villain as well as the hero of the Universe we know.

Man is more powerful than he is intelligent; and he interprets self-interest as something entailing short-term advantage rather than long-term survival.

Perhaps man is capable of no more than short-term advantage and that, perhaps, is the natural self-limiting aspect of the kind of power that outstrips wisdom. It leads, perhaps, to inevitable self-destruction so that the remnants of "lower" life can then limp on over what is left of his failure. The survivors can then proceed, without him, to a new and different (or better?) kind of life-flourish in the future.

This is one of a number of articles I have written in an attempt to help induce understanding of our will-to-suicide. A vain attempt, perhaps, but doing what little I can, helps me to sleep nights.

6

MAN THE OVERBALANCER

If we are truly to understand the grave human population threat that now faces us, we must first know something about the history of life on Earth and how we got where we are today. As surely as we live in the present, so are we equally the products of all that has gone before.

As evidence of our current plight, let me cite some provocative statistics. World population, if unchecked, will double Man's numbers in thirty years, says the U. S. Census Bureau. China's population will be about six billion by 2070 and that of America 420 million, unless curtailed. By 2100, the world total could be 25 billion.

The mind boggles at such figures, and we become even more confused when we try to analyze the factors responsible for our predicament to determine what should be done about it. A host of unanswered questions arise: When, why, and how did population increase get to be such a problem? Is it really as serious as we are told? Is overbreeding an innate characteristic of the human species? What, if anything, can science do to resolve all this? If science doesn't hold the answers, what does?

With these and related questions in mind, let us try to shed some light on the whole thicket of confusion. To do so, we must really begin at the beginning—in those misted eons when scientists believe the earliest stirrings of life probably first appeared.

The oldest known rocks with any appreciable fossil record are those of the Cambrian period approximately 600 million years ago. Life had probably existed for several *billion* years before that in microscopic form, but it is only in the Cambrian era that we first see signs of substantial, though primitive, organisms. Thus, it seems logical to begin our discussion at that point in time.

In the Cambrian period, all life forms, of which trilobites are the most typical, live in the sea; all are invertebrate. Life is listless; food consists of inanimate particles in the water; there are no predators. By the Silurian period, vertebrates—a new kind of creature with an internal skeleton, combining strength and mobility—have appeared and are now abundant. But these early vertebrates are relatively simple fishlike creatures, showing only the beginnings of progress. The first land plants have appeared, too, and the great thrust of life onto the land is about to begin. For over 90 per cent of Earth's history, the planet's surface has remained sterile and dead, but now plant life is creeping past the high-tide line.

Animal life follows in the Devonian period. Spiders, snails, and insectlike creatures live on the land plants. Fish with stubby fins and goggling eyes hobble onto the land, there to find other water in pools. Amphibians develop the capacity to live on land for extended periods, during adulthood at least. Special eggs evolve which can be hatched on land, and in the Carboniferous period, animals become capable of living on land their entire lives. This period also sees magnificent forests of fernlike plants that are to give rise to the coal beds of modern time.

Land-living reptiles flourish in the Permian and Triassic periods, and as they grow larger and more specialized, they proliferate in many directions of genetic specialty. Some of these reptiles later return to the sea; others develop long, webbed fingers and wings. Certain Triassic reptiles grow huge and subsequently become dinosaurs, the largest land animals ever to live on earth. At about the same time, some small reptiles are growing hair and developing warm blood to become the first tiny mammals.

In the Jurassic period, lizardlike reptiles grow feathers and they also develop warm blood, eventually becoming a new kind of flying creature, the bird. In the Cretaceous period that follows, reptiles have their last burst of vigor, reaching maximum size, and then most die out. By the end of this period, the dinosaurs, are gone, and it is the birds and mammals that will dominate hereafter.

In the Tertiary period, the mammals develop a high rate of metabolism which enables them to function in climatic extremes, and their brain grows more complicated. The primates, in particular, develop a brain large in comparison to their body size, with a notably wrinkled surface which presumably allows the presence of additional gray cells. Somehow about this time, perhaps not long before the dawn of the Pleistocene epoch a million years ago, some of the great apes develop into the two-legged hominids who are the ancestors of modern Man.

As a result of all this incredibly complex activity, there are today millions of species of living things. Each exists in a dynamic balance with the environment around it. When that environment is altered, species adapt to it in a process we call organic evolution. This dynamic process occurs continually, but when the environment is changed too suddenly or too forcibly, the species involved are subjected to more change than they can cope with.

Within the framework of the environment, each individual depends for a comfortable living on other individuals of his own or other species. Only the very simplest forms of life could live on an Earth that contained nothing alive but themselves. What's more, all life depends on the nonliving background of the environment that envelops it. Life depends on the air, the water, the soil. Without this vital nonliving background, no form of life as we know it could exist.

It is this interdependence of individuals and species and the dependence of life upon nonlife that are the concern of ecology. Earth's environment represents a complex, dynamic balance that is always in flux. The individuals of one species eat those of another, and each species benefits. The eater is fed and the eaten is eliminated.

Every individual, every species naturally tries to eat and to avoid being eaten. If any one species or group of species is unusually successful in this attempt, its numbers increase at the expense of the rest until Nature strikes the victors down and restores the balance.

Evolutionary or environmental factors may in the past have given one species a big survival advantage over others. But what we are faced with today is a situation created by the deliberate action of the first species in history, Man, that has proved intelligent enough to build a technology that could wreak havoc with the environment.

Even as a primitive hunter and food-gatherer long ago, Man showed signs of becoming a threat to the orderly structure of the environment. He grew intelligent enough to develop speech so that he could hunt and live in an unprecedentedly flexible and cooperative society. He developed tools—beginning with sticks and bones and going on to sharpened rocks—and these increased his power and flexibility.

There was nothing particularly new in this. Dogs and wolves hunted in packs, too; sharpened rocks were imitations of fangs and claws, and even spears and arrows, when these were developed, merely duplicated the work of taloned birds of prey. What was important was the *speed* with which Man developed these abilities. Where evolution slowly improved the efficiency of other species over millions of years, Man's intelligence changed and improved his own ways in mere thousands. Other species of comparable size could not possibly keep up.

Then, too, early Man made one technological advance that was unique—one before which no other creature could stand. He tamed fire. The warmth of fire helped him invade the colder regions of the world hitherto closed to him. The invention of cooking made food available that would ordinarily have been too coarse to chew or digest, and thus Man's diet was enhanced. And the campfire that cooked the food also kept prowling predators at a distance.

Man's increasing efficiency as a hunter, even while he was still savage, helped bring about the extinction of some of the very species he hunted—the mammoth, for instance. Man was only minimally destructive, however, as long as he remained a hunter and food-gatherer and few in number—perhaps ten million the world over. Animals could still outrun, outhide, or outbreed him well enough to survive.

Then, about ten thousand years ago, came the development of agriculture and herding. Man domesticated plants and animals. He deliberately raised and cultivated those which provided milk, eggs, wool, fibers, labor, food. This strained the normal balance of the environment in several ways. Farmland had to be irrigated, and as a result the very face of the earth was changed. (Animals admittedly affect the environment, too—the beaver and its dam building, for instance—but none on as devastating a scale as Man.) Man also altered Nature's balance by favoring the growth of certain plants and animals and by killing off competing species.

For thousands of years, the amount of land devoted to agriculture steadily expanded, and the number of wheat, barley, and cotton plants, to name but three, grew and grew and grew. Meanwhile, other plant life shrank correspondingly. The number of cattle, sheep, goats, pigs, horses, etc., grew, and the range and number of other animals declined.

With an assured food supply, humanity grew in numbers and had the leisure to develop further arts and technology—pottery, basketry, textiles, bricks, metals, cities, writing, science. By 1800, there were 900 million people on Earth, and the planet was beginning to show the marks of Man's deliberate use. Forests were being cut down and prairies plowed; everywhere there had to be farms and pasturage if Man was to live.

Not all this activity ended in Man's favor, of course. When an immensely complex environment begins to be altered, there are bound to be side effects difficult to prevent or even foresee. When cotton plants were growing wild in sparse clumps, for instance, some were parasitized by insects, but the insects that fed on them had a limited food supply since they experienced difficulty in getting from clump to clump. In large cotton fields, however, insects found a tremendous food supply, with one cotton plant almost on top of the next.

As Man farmed and herded all over the world, therefore, certain species of insects and rats multiplied with him and plagued him unbearably. Man, in his ever-crowding numbers, became an increasingly good mark for fleas, lice, and bacteria. Contagion grew easier and worldwide epidemics became common.

Yet Man survived, and the new factors he introduced into the world ecosystem seemed on the whole to favor him. Eventually he began to tap the great inanimate sources of energy. He began to use fire to heat water in an enclosed space and let the expanding steam perform useful tasks.

By 1800, the steam engine was beginning to introduce important new changes in society, and Mankind in turn had become such an influential factor in the whole fabric of life that almost any innovation of his, however small, altered that life fabric, often in a large way. With the advent of the steam engine and the Industrial Revolution, Mankind built up a vast technology in the space of scarcely two centuries. Transportation was so dramatically improved that food could be transferred easily from place to place, and the mechanization of agriculture and the use of chemical fertilizers increased the amount of food so transported, thus reducing famine.

The birth of modern medicine, the introduction of effective methods of hygiene, the chlorination of water, the development of insecticides and antibiotics, all combined to defeat diseases which man had previously been helpless to combat. The death rate thereupon fell precipitously. As life expectancy rose, so did the population. By 1971, world population was 3,600,000,000, four times its size in 1800, and it is still increasing at a record pace, as noted earlier.

The steadily increasing rate of industrialization since 1800 has been changing the inanimate background of life, too. First coal, then oil and gas have been burned in rapidly increasing quantities to supply the energy needs of ever-more human beings demanding an ever-higher standard of living.

The result has been a flood of soot and impurities that have fouled the atmosphere, while chemical wastes poison the waters and garbage and rubbish piles up in every corner of the planet. The introduction of nuclear fission has recently added the new problem of radioactive waste disposal.

Then, too, mineral wealth is being extracted from the Earth's crust at a steadily increasing rate to support the technology that supports the ever-higher living standard. To be sure, all these resources are eventually returned to the earth, but that does not help. Minerals are withdrawn from concentrated pockets—where the concentration has been accomplished by slow geologic change over millions of years—but the minerals are returned in a thin, mixed scattering that makes them extremely difficult to reconcentrate.

And always there are more people—cities and suburbs spread, new towns are born and grow; and houses, houses, and more houses are built, encroaching on the wilderness in all directions and everywhere. More people means more animals to supply Man's needs, more plants for himself and for his animals, less room for all other living creatures.

There are, of course, creatures that have adapted to the manmade environment and thrive in it; urban rats for instance. (Their number now equals that of the human population in American cities.) Also, there are the insects, bacteria, and viruses that show no signs as yet of retreating. They multiply so quickly and live, as individuals, so briefly that their rate of evolutionary change is fast enough to match the changes mankind produces.

While this transpires, the friction of man against man increases. The crowded cities fester; society, growing larger and more complex each year, becomes top-heavy and unstable. The chances of social unrest, civil strife, and international war increase.

Some people deny that the situation today really differs much from the past. There has always been pollution, they say, and the crowded, medieval cities were rampant plague spots. (London today is admittedly less smoggy than it once was.)

Ah, but even if particular places have improved, the situation as a whole is racing downhill. The overloading of the world ecosystem is progressing at an ever-faster rate, and it is no comfort to find a few places where the line is temporarily being held.

Can it be argued, for example, that the world is still underpopulated just because nations such as Canada need not presently be concerned about their birthrate? Well, what about the Netherlands, you might ask. It has an extremely heavy population density, and yet it is a comfortable, beautiful, and civilized nation. The Netherlands, like the rest of the industrialized countries, depends for its raw materials on areas of the world where, ironically, living standards are far lower than its own. In fact one can argue that the Netherlands is as well off as it is *precisely because* so much of the world, though rich in resources, has such a low living standard. If all the world were as industrialized—and as crowded—as the Netherlands, we would dangerously deplete the resources we now take so much for granted.

Well, then, where will it all end? What about the recently published report that world population, if unchecked, could double Man's numbers in the next thirty years? Are these just interesting speculations, or do they hold within themselves the greatest danger Mankind has ever had to face? Let's see....

It has been estimated that the total weight of plant life on Earth is twenty million million tons. This mass depends on the energy of sunlight. Only so much sunlight reaches Earth, and only a small fraction of that can be used by plant cells. Hence, to increase the total quantity of plant life substantially would be extremely difficult.

Yet all animal life depends on plant life for food. (Some animals eat other animals which eat still other animals, but eventually the food chain ends with some animal that eats plants.) As a general rule, there is a ten-to-one weight ratio between that which is eaten and the eater. Thus, the twenty million million tons of plant life on Earth can support about two million million tons of animal life.

In instances where animal life increases beyond normal limits, plant life is eaten faster than it can replace itself. The total food supply then declines, and animal life dies off through starvation until the balance is reestablished.

Suppose we estimate the average weight of a human being (counting both adults and children) to be 100 pounds. The total weight of the 3,600,000,000 people now alive comes to 180 million tons. That means that presently about 1/100 of 1 per cent of the supportable two million million tons of animal life on Earth is human. That doesn't sound like much, but probably never in all the Earth's history has one species of animal life ever formed so large a percentage of the total weight.

If Mankind's present rate of population increase continued for another three hundred years, Man would make up about 10 per cent of the total weight of animal life on Earth. The animals Man feeds on and uses for various purposes would then make up almost all the rest. Wildlife would be just about totally wiped out.

In a little over four hundred years, the weight of humanity would equal the total present weight of all animal life, and the human population density would be much higher on the average, *all over the world*, than it now is on Manhattan Island alone. Can you picture an Earth consisting of one huge structural complex, both residential and industrial, covering the entire global surface, both land and sea? Can you picture a world roof consisting of an immense ocean of algae growing in sunlight, in order to supply food and oxygen for Earth's vast human population? In that world, there would have to be a forced ecological balance consisting of a single animal species, Man, and his food. Do we really want an Earth made up of little but men and algae?

If not, how is an ecological balance to be maintained? Unless the advance of science is rapid and human social organization perfect, starvation, epidemics, and social unrest will send the death rate sky-high and end the population explosion. The only reasonable alternative would seem to be to end the population rise by lowering the birthrate. But how? There is a strong drive, both biological and social, to have children. There are intense controversies over methods for lowering the birth rate, and even over the value of doing so by whatever means.

What to do? Does science hold the answer? Can the so-called "new biology" be turned from the medical concerns that now occupy its primary attention to the specific matter of the birthrate?

There is a pleasure center in the brain which, when stimulated electrically, gives rise to sensations of ecstasy. All the ordinary pleasures of life seem to us to give pleasure only to the extent to which they activate that center in our brains. Might it be, then, that everyone could be outfitted with a device allowing him to activate his own pleasure center at will? Might he not then be convinced that this should replace the lesser pleasures of sex? Might that not in turn cause the birthrate to drop—even to zero?

Such a solution would undoubtedly bring about new problems as bad or worse than the old. If people could tap their own pleasure centers, would they want anything else? Of what value would they find the lesser, more indirect pleasures they now derive from artistic creation or scientific research, or from satisfying the itch of curiosity and ambition? If we must destroy character to save Man, what have we saved?

Perhaps it is not a matter of adjusting people physically but merely psychologically. The Harvard psychologist B. F. Skinner believes that normal men and women are almost entirely products of the environmental influences around them. No one, in Skinner's opinion, does as he chooses, but only what his environment dictates.

If Skinner is right, it would merely be necessary to adjust the environment in such a way that individuals would act as they ought to act in an overcrowded society. We would simply push those particular environmental buttons that would make men less anxious to have children, more eager to refrain from polluting, more considerate of wildlife, more anxious to think of the group before themselves.

But would this work? In the first place, do human beings really function as Skinner thinks they do? Not all psychologists believe so by any means. And even if Skinner is right, who would decide the type of behavior best designed to solve Man's troubles? And who would set up the buttons, and who would push them?

And wouldn't there need to be a special environment designed to develop button-pushers with the proper attitude and ability to push buttons? And who would push the buttons to develop the proper button-pushers?

Obviously, we can't hope for glib, one-shot solutions. There must be contributions from every direction. Scientists must develop new methods of birth control and new understanding of the human brain; psychologists must develop new techniques for directing human behavior; conservationists must devise new methods for preserving the environment, and sociologists and statesmen must develop new institutions for preventing war.

Most of all, though, we must depend on the good sense of people generally—heightened by the gathering misery—to adopt a new attitude toward childbearing, and to make a new effort to think in terms, not just of themselves, but of the entire Family of Man. Perhaps everything together can combine to edge individuals toward a clearer consciousness of the manner in which their own safety and comfort is bound up to all Mankind and, beyond that, to the total environment.

PART TWO

LIFE PRESENT

I was asked to write an article in honor of the thirtieth anniversary of the founding of the United Nations and I did so, although I found myself filled with sad disappointment. I am old enough to remember the hope and ideals with which the United Nations was founded. It was to be an organization that, unlike the older defunct League of Nations, was to rise superior to destructive nationalism and to be a mouthpiece for mankind united.

The very name indicated that. The new organization was not to be a mere "league" of independent self-seeking nations; it was to be those same nations "united" in a common search for a common goal. Alas, the stupidity of man seems, so far, to be unconquerable. The United Nations has become merely a rather despicable forum used for private nationalistic ambitions, with each nation forming shifting alliances to see which can have the honor of best hastening mankind's destruction.

And yet—I suppose hostile talk is better than hostile acts; and the United Nations offers a forum which, however unworthily it is now so frequently used, may be better used in the future.

So I wrote the article that follows, taking care to emphasize my own globocentric view.

7

THE MYTH OF LESS-THAN-ALL

The history of civilization has been the tale of a slow widening of the ripples. Each century, a military or political upset in one place has made its effects felt at a greater and greater distance from that place. Each century, a particular society has grown less and less able to ignore the turmoil elsewhere, and the insulating distance to insure indifference has grown longer. For example:

In 1650 B.C., it did not concern the Greeks that the Egyptian Middle Kingdom, five hundred miles away, had fallen to the Hyksos invaders. In 525 B.C., however, the fall of Egypt to Persia so clearly heightened the dangers to Greece that the land recognized it was in a crisis. Greece was never again in its history to be unaffected by events in the eastern Mediterranean.

In 215 B.C., the deadly duel between Rome and Carthage raised no echo in the hearts of Britons on their tight little isle, one thousand miles away. In A.D. 407, however, the state of Italy with respect to invaders was of intense interest to Britain, for Alaric's presence in northern Italy cost Britain its Roman garrison and, temporarily, its civilization. Never again in its history was the island to be unaffected by events in Western Europe.

Even as late as A.D. 1935, most Americans could still live as though it didn't matter what happened in Europe, three thousand miles away. In only thirty more years, however, Americans would be told, and would believe, that what happened in Saigon, ten thousand miles away, was of such vital importance to Kansas that tens of thousands of Americans must die.

The United States cannot remain indifferent to turmoil anywhere in the world any longer. Nor can any other nation.

To suppose that any group of people need concern itself only about itself and its immediate neighbors is to live in a dream-world. That is not the way things work anymore. Any feeling that less-than-all will do is a myth.

The widening ripples set up by mankind through history have been the result of the advance of technology.

Advancing technology has extended the reach of various societies, making it possible for them to find their resources at greater and greater distances from home, and has increased their needs and appetites for those resources as well.

Now the reach has grown worldwide. Now all the world competes for all the world's resources. No nation, however large, however populous, however wealthy, however advanced, can any longer support its numbers and its complexity and its unlimited ambitions by making use of only the land, sea and air within its own political boundaries. Each nation needs the others, and is needed by the others.

There are nations rich in some of the requirements for a prosperous society and poor in others. There are also nations poor in all the requirements for a prosperous society. There are, however, no nations rich in all the requirements for a self-sustaining society.

It is only *the whole word as a unit* that is, as yet, rich in all respects (provided we limit our numbers and grow wiser in the use of our power). Anything less-than-all? Forget it. That is a myth.

Mankind has major problems. Most of them can be traced to our technological advances, but they are the side effects of the benefits we have received.

The thought that we might solve our problems now by abandoning technology is a wildly foolish one. Abandoning technology isn't really possible, and no one really wants it, not even those who believe they want a return to a simpler life.

In minor respects, such as the abandoning of electric toothbrushes or foregoing automatic car-window controls to cut down on energy waste or in regulating the manner in which we handle wastes, there can be small retreats, but only small ones.

Consider, for instance, the overriding problem of humanity—its wildly growing population. Consider mankind's increasing numbers that are outstripping the world's food supply, outracing its energy resources, outgrowing its room, outraging its ecology. The origins of this problem are the development of the germ theory of disease in the 1860s and the manner in which medical science then proceeded to score new victories over disease each decade thereafter. Rapidly, the death rate was lowered over more and more of the world, while the birthrate was left high enough to allow the yearly rise in population increase to the present record figure of 2 percent a year—200,000 additional mouths *each day*.

Shall we, then, abandon our medical science and allow pestilence to carry off its hundreds of millions and, in this way, thin out our numbers? Which of us can be sure we will survive, and which of us would enjoy surviving in a world sinking into the chaos of unbridled plague? Surely the better alternative is to keep our advanced medical science and use it to devise methods of controlling the birthrate as well as the death rate—and so on in every other problem we face.

Mankind has only a forward gear that will operate in safety, no other. To go into reverse is sheer, unimaginable catastrophe. Even if we reach the point where the forward gear will also lead to catastrophe, that will not make the reverse gear any safer. It may be that, in the end, there is no escape at all, but if there should be escape, it can lie in only one direction: Forward. There must be still further advances in technology; advances, it is to be hoped, more wisely managed than some we have had in the past.

If these advances bring problems of their own, that is the nature of the universe, and we have no choice but to continue forward to solve those problems in their turn by further technological advance—and then to solve the problems newly arising—and then the next—and so on.

If that seems a boring, endless and unrewarding task, then please consider the catastrophic alternative.

Think of the various problems that face mankind today—unrestrained population growth, shrinking resources, increasing pollution, a withering ecology, suffocating military expenditure, intensifying violence and, in every respect, the sad signs of a society growing psychotic.

All of these have one thing in common—they affect all mankind and are therefore not amenable to local solutions.

When technology supplies the solutions, they must be applied on a worldwide basis with full worldwide cooperation, if they are, indeed, to be solutions.

No reduction of the birthrate through the proper use of chemical or mechanical controls, through the use of economic rewards, social pressure or educational enlightenment will be worth anything if it is not applied over all the world.

To increase the food supply by an ordered, systematic and careful culling of the oceans, by the development of new strains of grain, by the more efficient distribution of fertilizer, won't work if all is not carried through on a worldwide basis.

If the stranglehold of wasteful military expenditure and the fatal sword of war is to be removed from the throat of humanity, can it be removed from one group of nations while others maintain the threat?

In a world that has grown utterly interdependent, there can be no islands of safety and sanity. Where a highly industrialized society needs the resources of the world, it cannot maintain itself if the whole world cannot maintain itself. Safety for anything less-than-all is a myth.

If we are to look to technology to solve these problems, again we must broaden our field of vision. The days when any nation—or any small group of nations—had a monopoly on science are over and can't return. The increasing complexity of our growing knowledge of the universe makes it necessary to use the entire pool of humanity as a brainpower and information source. Nothing less-than-all will give us enough.

All the world represents the brain supply for all the world. All the world represents the resource well and the waste sink for all the world. All the world suffers the various *world* problems and must be part of the various *world* solutions.

In the 1940s, the nuclear bomb was developed by something that Americans jubilantly called "Yankee know-how." Actually, it was developed by the united efforts of scientists from a dozen nations or more. To recall the role of some of the notables—Fermi, Teller, Szilard, Einstein, Bohr, Frank, Chadwick—is to call the roll of Europe as well.

Since then, the world has had to cooperate in projects that are global in nature. Antarctica has been, and is being, explored on an international basis. The world's weather is a matter of international concern, and the World Weather Watch the UN is running is absorbing information from every corner of the world that will be valuable to every nation in the world.

How do we farm the sea and how do we mine the sea bottom? How do we tap the inner heat of the earth or tame the tides? How do we cure cancer? How do we end hunger? Every important question today calls for the maximum efforts of scientists everywhere.

The greatest technological prize awaiting us today—that of finding a way to place controlled fusion energy at the service of mankind—must be a cooperative world effort, with scientists in the United States, Great Britain, the Soviet Union and elsewhere freely trading knowledge among themselves.

Is there anyone hardy enough to suppose that the great technological prize of the twenty-first century—the colonization of space and the manned exploration of the solar system—can possibly involve anything other than a global effort?

To suppose that, at this present stage of social, industrial and informational complexity, further technological progress can be carried forward on a less-than-all basis is to fall prey to a myth.

We live on a small ball of rock that is all one piece. Anything less-than-all is a myth.

Yet we inherit a nation-state system from the nineteenth

century and before. Almost all of us are dedicated, somehow, to the belief that the needs and desires of our own nation are of greater importance by far than those of any other. Our "national security" (that's the phrase) must be guarded by bristling weapons and resolute men and, if necessary, protected by unrestrained violence. All damage everywhere in the world is justified as long as our own particular corner of the world can be said to benefit.

But that, too, is a myth. There is no way in which a nation's security can be preserved any longer by anything less than preserving the security of all mankind. The very effort to protect a nation, one small portion of mankind, by armed might diverts brains and resources from the effort to solve the world's problems. It makes less likely the preservation of the security of all mankind and, therefore, the security of any nation making up mankind.

Here, too, salvation lies only in giving up the less-than-all myth. It is the only sane choice.

But will mankind adopt the sane choice? We are not compelled to. There is always the alternative of choosing the insane course of continuing on the present route and ending in absolute catastrophe in perhaps no more than thirty years.

If the insane choice is the one chosen (and it is only too likely to be), it is not because people want catastrophe. It is because they cannot shake themselves free of the myth of less-than-all long enough to see that all mankind is the smallest viable unit on earth.

If, on the other hand, the sane choice, against all likelihood, is chosen, it means that the nation-states who now hold the allegiance of the world's peoples, and who face each other with the perpetual threat of war, must learn to cooperate so fully, intimately and genuinely that we will have what will be, in all essentials, a world government.

It is sad that something so essential to survival as world government should arouse such stubbornly adverse thoughts in the minds of those who must surely long to survive. It is as though they see in a world government a device for being forced to give up all one's cherished ways of life at the behest of a bunch of "foreigners."

Well, think about it! Many of our cherished ways of life will have to be modified. A falling birthrate and a food shortage may well leave us with completely different attitudes toward "Mom and apple pie." If survival is the name of the game, change is the language in which that name is written. And if it is any comfort to you, not only will you have to change your ways, but so will all those "foreigners."

No doubt, we can use our language to help us past the shock. We can save our sensibilities by speaking of "international cooperation," or "a multinational dialogue," or a "global emergency conference." It doesn't really matter what it is *called*, as long as it *is* a way of governing the world as a whole.

Fortunately, we have a beginning in the United Nations and, fortunately, that is something we are used to and therefore not too frightened of.

Born in the aftermath of World War II, the United Nations is the living response to the fact that our planet is too small to be separated into nation-states.

The United Nations lacks the power to force its decision on its member nations directly and often appears to be just a useless talking machine. However, it represents an idea.

The United Nations is the idea of a united concern for the problems and needs of mankind, the idea of a concerted road to safety.

It may graduate to something better if the myth of less-than-all evaporates. It may become the nucleus of a world organization that will gather together the arms and brains of humanity in an attack on the world's problems and then direct the careful implementation of the world's solutions. Indeed, the necessary cooperation of the world's scientists and of the nations supporting those scientists, in working on such clearly international problems as the environment, population and epidemic control, may well serve as the prototype for deeper, broader, and more permanent international cooperation on other kinds of problems.

In this way the UN may serve to keep mankind secure and to arrange a new society in the twenty-first century that will live within the limits of the world's resources and that will then strike outward to new horizons in worlds beyond earth.

Or it may not, and we will be destroyed.

The choice is ours and, for what it's worth, we need not wait long. If we do not stir ourselves to choose sanity and life within the next thirty years at most, we will have succeeded in choosing the alternative of insanity and death.

As much a part of life as man's physiological past is his intellectual past, his discovery of knowledge. So recent has much of this discovery been, not only as compared to the age of terrestrial life, but to the age of man as a species, that it seems fair to consider such discovery as part of life's present.

The first aspect of this discovery is mythological. Myth seems to be an unsophisticated way of looking at the Universe, at least from our present vantage point, but it was nonetheless a real attempt to understand the Universe. It is the attempt, quite apart from what success it may gain, that is the measure of the dignity and wonder of the human mind.

8

THE FLAMING GOD

If you were a primitive person waiting through a long night; if it were dark and chilly, with no source of light and heat but perhaps a smoldering campfire; if you could hear out there the rustling noises that might mean predatory animals who could see far better in the dark than you could; if you could sleep no more—what would be the greatest sight in the world?

It would have to be the soft graying of the sky in the east, the brightening of the dawn that brought the sure promise that, in a short while, poking above the horizon, would come the Sun itself, to make the whole world light and warm and secure again.

In those days, when the workings of the Universe were attributed to myriad gods, surely among the chief of them would have to be a Sun-god, powerful and beneficent, for how could human beings live without the Sun. Even in the Bible, God's first command was "Let there be light" (to be collected into Sun, Moon, and stars on the fourth day), for without light nothing else was possible.

To the ancient Egyptians, the Sun-god was Re, and he was

the very principle of creation, having created not only everything else, but having even created himself. Each Egyptian city had its own god, often equated with the Sun-god. When the Egyptian Empire was at its height about 1500 B.C., with its capital at the southern city of Thebes, the god of that city, Amon, became Amon-Re, god of Thebes and the Sun.

Then, later, when, for the first time in history as far as we know, a monotheistic belief was established, briefly, under the Pharaoh Ikhnaton of Egypt about 1360 B.C., the one supreme god he worshiped was the god of the Sun.

The equally old Babylonian civilizations had a Sun-god they called Shamash, the giver of life and light, and the father of law and justice. And why not? It's natural to equate law and justice with the light of the Sun and to feel that the cloak of darkness hides evil and crime. Even today the streets and parks of American cities seem to be abandoned to dubious denizens by night while the honest citizens creep out to take possession only in the light of day.

Every civilization had its Sun-god among the great powers of the pantheon. India had the redheaded Surya, from whom the race of human beings was descended. Japan had Amaterasu (unusual in being a Sun-goddess, female rather than male), and if she was not the ancestress of the human species, she was at least the progenitor of the Japanese ruling house, of whom Hirohito is the current representative.

The Norse had the beautiful Balder, god of the Sun, of youth, and of beauty, who was married to Nanna, the goddess of the Moon. —And so it goes. The ancient Irish had Lugh; the ancient Britons, Llew; the ancient Slavs, Dazhbog (who was also the god of wealth and success—undoubtedly because of the Sun's golden appearance); the Polynesians had Tane, who was also the god of all living things; the Mayas had Itzamna, another Sun-god who was the first, the oldest, and the creator of all else; the Aztecs had Quetzalcoatl, a Sun-god who was also the god of wisdom and who invented the calendar.

The best-known Sun-god to us of the Western tradition is, however, the Greek Helios, who, in later Greek poetry, was identified with Apollo. Where the Egyptian Sun-god, Re, crossed the sky in a boat (the typical Egyptian transportation device along the River Nile), Helios crossed it in a magnificent golden chariot drawn by four fiery horses that only he could control.

The difficulty of keeping those raging steeds in their course was the thought that gave rise to perhaps the best-known myth involving the Sun-god in Western literature. Helios had a son. Phaëthon, by a mortal woman. When doubts were cast on his paternity, Phaëthon went to Helios and demanded the god vow to vindicate his son's honor. Helios vowed and Phaëthon demanded to be put in charge of the solar chariot for a day.

Helios was forced to give in and Phaëthon took the controls. Feeling an inept hand on the reins, the horses went out of control. Rearing and plunging, they came too close to Earth, burnt a desert across northern Africa and baked the African peoples black. The Earth would have been destroyed if the Greek master-god, Zeus, had not struck Phaëthon out of the chariot with lightning and allowed the horses to return of their own accord to their accustomed path.

The normal path of the Sun can itself be interpreted as a matter of adventure. To help use the Sun and Moon as bases for timekeeping, the ancient Sumerians (the earliest civilization in the Tigris-Euphrates valley) were the first to mark off the stars into those groups we now call constellations and give them fanciful names based on distant resemblances of the star configurations to familiar objects. The Sun, in the course of the year, passed through twelve constellations of the zodiac, called after the names of lions, crabs, archers, and so on.

The tale of the Sun's journey would recount his victory over each danger he encountered, and the suspense would be great, for only by his victory could his course be successfully completed and human survival assured. It may be that the twelve labors which Hercules must successfully complete before achieving rest in heaven is a version of the Sun's passage through the twelve dangerous constellations—a version obscured by changes in the names of the constellations and by the endless accretions of incidents by the mythmakers of ancient times.

Yet the Sun's career is not one of unalloyed success. However triumphant he may be ordinarily, he can be obscured by clouds. In those European lands where clouds and storms are common, it may be the lightning-wielding god of the sky or of storms who is supreme—the Zeus of the Greeks and the Thor of the Norse. Even the Bible seems to depict Yahveh as having been a storm-god in primitive times.

There is, then, the danger of eclipse that temporarily seems to slay, in part or in whole, either Sun or Moon. In the Norse myths, both Sun and Moon are eternally pursued by gigantic wolves as they make their course across the sky, and occasionally the wolves overtake the luminaries and hide them, temporarily, within their slavering jaws.

But the storm cloud is occasional and the eclipse is even more occasional. One Solar death, however, is regularly periodic and inevitable. At the close of each day, the Sun, no matter how glorious its reign, must sink beneath the western horizon, defeated and bloody, and Night returns in victory to the sky.

This is represented most colorfully in the Norse tale of the Sun-god, Balder. Balder, the joy of gods and humanity, is troubled, suddenly, by a presentiment of death. His mother, Frigg (the wife of the Norse master-god, Odin), exacts an oath from all things to do no harm to Balder but neglects the mistletoe. The gods then engage in the game of hurling missiles at Balder in order to watch the missiles swerve away of their own accord.

The evil god of fire, Loki, learning of the exemption of the mistletoe, carves a mistletoe branch into a spear and gives it to Hoder, the god of Night, who, being blind (after all, one cannot see by night), does not participate in the game. Loki guides the aim and Balder falls. The Sun has died under the attack of Night.

A less obvious solar myth may be the Hebrew legend of Samson. The Hebrew version of the name, Shimshon, bears a striking resemblance to "shemesh" the Hebrew word for the Sun (itself related to the Babylonian, Shamash). Two miles south of Samson's traditional birthplace was the town of Beth-shemesh ("house of the sun") believed to have been a center of Sun worship.

Samson, like Hercules, survives various dangers, thanks to his superhuman strength. What's more, Samson's strength lies, explicitly in his hair, which may be viewed as representing the golden rays of the midday Sun. When Samson's hair is shorn, he grows weak, as does the Sun, when it approaches the horizon, red and rayless, so that it can be looked on without harm. It is in the lap of Delilah that Samson sleeps when he is shorn and Delilah's name is closely akin to the Hebrew "lilah" meaning "Night." The Sun sinks into the lap of Night and is defeated and blinded. But Samson's hair grows again and he recovers his strength for one last feat since, after all, the Sun does rise the next morning.

In fact, in the sunny lands particularly, the Sun must survive all the onslaughts of night and win in the end. In Persian mythology, Ahura Mazda, the god of light, fights Ahriman, the god of darkness, in a cosmic battle that fills the Universe—and it is Ahura Mazda who will win at the end of time. (The Jews of the Persian period picked up this view and it is from 400 B.C. onward that Satan enters the Judaic, and later the Christian, consciousness as the dark adversary of God, to be defeated at the end.)

The Sun's setting and rising is one inspiration for the many mythic tales involving the death and resurrection of a god. An even more impressive death and resurrection is the death of vegetation with the coming of winter and its restoration in the spring.

The tale of Balder might just as well be the symbol of the god of Summer being slain by the god of Winter. Similar significance can be given to the death and resurrection of Osiris among the Egyptians, of Thammus among the Babylonians, of Proserpina among the Greeks, and so on.

But the Sun is clearly connected with the summer-winter cycle as well as with the day-night cycle. Throughout the European summer, the noonday Sun reaches a slightly lower point in the southern sky each day than it did the day before. As the Sun's path in the sky slowly sinks southward, the temperature grows colder and the vegetation browns and dies.

If the Sun should continue to sink, and should pass down behind the southern horizon altogether, death would be universal and permanent, but that does not happen. The rate of sinking slows and each year at December 21, by our calendar, the Sun comes to a halt ("solstice"—or "Sun halt" in Latin) and thereafter begins to rise again.

The winter may continue to sharpen after the solstice but the fact that the noonday Sun is rising steadily higher in the sky is a guarantee that spring and summer will come once more. The day of the winter solstice, of the birth of a new summer Sun, is therefore a time for a great festival, celebrating the rescue of all life.

The most familiar solstice celebration of ancient times was that of the Romans. The Roman agricultural god, Saturn, was believed by them to have ruled Italy during an early golden age of rich crops and plentiful food. The winter solstice, then, with its promise of a return of summer and of the golden time of Saturnian agriculture, was celebrated with a week-long

"Saturnalia" from December 17 to 24. It was a time of unrelieved merriment and joy. Businesses closed so that nothing would interfere with the celebration and gifts were given all round. It was a time of the brotherhood of humanity, for on that day servants and slaves were given their temporary freedom and allowed to join in the celebration with their masters and even to be waited on.

The Saturnalia did not disappear. In fact, other evidences of Sun worship came in the time of the later Roman Empire. Heliogabalus, a priest of the Syrian Sun-god, sat on the Roman throne from 218 to 222, and about that time, the worship of Mithras, a Sun-god of Persia, was becoming popular, especially among the soldiers.

The Mithraists celebrated the birth of Mithras, the Sun, at the winter solstice, a natural time, and fixed the day on December 25, so that the popular Roman Saturnalia could build up to the Mithraist "Day of the Sun" as a climax.

At that time, Christianity was locked in a great duel with the Mithraists for the hearts and minds of the people of the Roman Empire. Christianity had the great advantage of accepting women into the religion while Mithraism remained stag (and, after all, it was the mother, not the father, who influenced the religious belief of the children). Mithraism, however, had the Saturnalian festival of the Sun on its side.

Some time after 300 A.D., Christianity managed the final coup of absorbing the Saturnalia and, with that, it scored its final victory over Mithraism. The birth of Jesus was fixed on December 25, and the great festival was made Christian. There is absolutely no biblical authority for December 25 as the day of the Nativity; in fact, from the biblical tale one can be quite certain the Nativity came at some other time, for there would be no shepherds tending their sheep on frozen fields regardless of what the Christmas carols say.

All the appurtenances of the Saturnalia were adopted anyway—the joy and merriment, the closed businesses, the brotherhood, the gift-giving. All was given new meaning, but all was still there.

So that underneath the panoply of celebration of the birth of the Son is the distant echo of that far older rite, the celebration of the birth of the Sun.

This is another essay written as a TV Guide *backgrounder.*

Since, with TV Guide, *I reach millions of people whom I never reach otherwise and who are not accustomed to my secularist viewpoint, I await letters with what stoicism I can muster. In the case of this essay, there were objections to my description of the Bible as lending credence to the theory that disease is caused by evil spirits.*

I find it wearisome to argue over the meaning of the Bible, but I have just leafed through it at random and here is the first verse I came across in this connection: "When the even was come, they brought unto him [Jesus] many that were possessed with devils: and he cast out the spirits with his word, and healed all that were sick." (Matthew 8:16.)

One can argue that the word "devils" and "spirits" need not be taken literally, that they fulfill the same purpose that our present words "germs" and "bacteria" do.

What does it matter, though, if our contemporary religious sophisticates squeeze out symbolism to mask their own embarrassment at a biblical view they consider primitive. Unfortunately most people have, in the past (and now, too, for that matter) interpreted such verses literally, acted upon those interpretations, and brought untold misery to the world.

It is not so much being wrong that does the damage, as clinging to man's error by calling it God's truth. If there is an unforgivable blasphemy, surely that is it.

9

BEFORE BACTERIA

In the days before the scientific view of the Universe became current, it was easy to believe that all the mysterious phenomena about us were the work of invisible, supernatural beings—good, evil, and indifferent. Among the evil spirits were those which visited calamities on human beings: which took over human bodies, for instance, and caused disease.

How does one get rid of the evil spirits and cure the disease? By magical incantations, for one thing. By noxious potions, for another, for these were designed to disgust the spirit and cause it to vacate the body.

The ancient Egyptians had already worked out elaborate methods, both magical and chemical, to combat the evil spirits, and this has continued right down into contemporary times. Folk remedies, even today, are full of weird spells and potions.

The Bible lent the evil-spirit theory of disease its authority since the Gospels carefully describe how Jesus cured disease by casting out devils. As a result, it was quite common in medieval and early modern times to mistreat the mentally ill in brutal fashion in an effort to force the evil spirits out of them. Even today we exploit the theory in such motion pictures as *The Exorcist*.

The first prominent movement away from the evil-spirit theory came in Greek times. About 400 B.C., Hippocrates and his followers suggested that disease was not an invasion from without, but a disorder within. The various substances of which the body was constructed, they said, had a proper balance in people who were well and an imbalance, in which one substance might be superabundant and another deficient, in sick people.

About 300 B.C., a Greek physician, Erasistratus, suspected that the chief cause of this imbalance within the body was a superfluity of blood. This began the notion of bleeding patients to cure them; something which continued for two thousand years and helped kill numerous individuals faster than the disease unaided would have done. As late as 1799, it helped kill George Washington of an illness from which he might possibly have recovered if the doctors had stayed away.

Another theory of disease placed the blame on the influence of the stars in evil combinations. This astrological theory has left its trace in the name of "influenza" which is the Italian word for "influence." Still another attributed disease to bad air and we still have the disease "malaria," a word which is Italian for "bad air."

In ancient times, oddly enough, no one seemed to notice that some diseases were contagious.

In the Bible, however, there were careful descriptions of the manner in which people who had various skin diseases (lumped together as "leprosy") were to be isolated from the general population. This was for religious reasons rather than out of fear of infection.

The instructions were followed in early medieval Europe, however, and because such isolation did seem to cut down the incidence of such skin diseases, the idea arose that isolation might be effective in other cases.

During the 1300s, therefore, the practice of "quarantine" arose. (The word is from the Italian word for "forty" since patients were isolated for forty days.) Quarantine did help stop the spread of a disease, while the absence of quarantine encouraged the spread, so people began to understand that diseases might be contagious. Then in the 1300s, when the Black Death struck with unprecedented fury, the fact of contagion, already in people's minds, became overwhelmingly self-evident.

Once contagion was understood, there was a growing reluctance to come into contact with ill people and with anything they touched, so that notions of hygiene began to arise. Such notions advanced only slowly, however.

As late as 1847, a Hungarian physician, Ignaz Semmelweis, failed in his attempt to force the doctors in a Viennese hospital to wash their hands before delivering babies. Semmelweis was driven out and hand-washing stopped. The number of women who died of childbed fever dropped drastically during the short interval in which doctors washed their hands and went right up again when the hand-washing stopped.

In 1546, an Italian physician named Girolamo Fracastoro published a book that represented the first thoughtful consideration of the process of contagion. After describing the various ways in which a disease could spread, he suggested that there might be tiny bodies, too small to see, which were present in ill people and which might make their way, by direct or indirect contact, to well people. In the well people, these minute bodies would multiply and make them ill, too.

Fracastoro happened to be right, but as long as the tiny bodies could not be seen or detected in any way, this did not represent a real advance over the evil-spirit theory.

In the 1670s, however, a Dutch lens-grinder, Anton van Leeuwenhoek, produced the first lens that was well made enough to magnify tiny objects without distortion. In 1677, he was able to see living creatures through his "microscope" that were too small to see with the unaided eye. Though less than 1/50 of an inch long, they lived and multiplied in drops of water.

In 1683, Leeuwenhoek just barely managed to make out still smaller objects that we now know must have been bacteria. It was not for another century, however, that microscopes became good enough to see bacteria in some detail. In 1786, in a book by the Danish biologist, Otto Friedrich Müller, bacteria were described and classified for the first time.

Were bacteria the tiny bodies Fracastoro had thought might exist?

For this to be so, specific types of bacteria must be detected in all people with a specific disease, but not in people lacking that disease. The development of a disease must be shown to be accompanied by the appearance of the bacteria.

This was demonstrated by the French chemist, Louis Pasteur, and the German physician, Robert Koch, in the 1860s and 1870s. With this "germ theory of disease" physicians began the conquest of contagion which, in a century, was to double human life expectancy from thirty-five years to seventy years.

In 1975, the Astronomical Society of the Pacific gave me the Dorothea Klumpke-Roberts Award "for outstanding contributions to better public understanding and appreciation of astronomy."

I was dreadfully flattered and pleased, of course, but some of the edge of self-satisfaction was removed by the news that in return for the award I was to write an article for the Society's magazine. To that magazine, professional astronomers contribute and the level of its printed material is high.

It took me a while to overcome something that was remarkably akin to stage fright, but then I managed to turn out the following essay, which the Society published without any sign of discontent, thank goodness.

10

THE FACE OF THE MOON

It may be that the most important astronomical phenomenon in the heavens is the completely accidental one of Earth's possessing a satellite so large and so situated that we can see its face in some detail with the unaided eye.

That this is accidental is clear. Venus, which, in size, is nearly Earth's twin, has no satellite at all. In fact, if the planets of the Solar system other than the Earth are considered, those satellites which exist possess masses that are but small fractions of that of their primaries. If, in light of that, we were asked to guess the size of Earth's satellite without prior knowledge of the actual situation, we would suppose it to be no more than a couple of hundred kilometers in diameter at most, if it existed at all.

But, as a matter of fact, our satellite, the Moon, is 3,475 kilometers (2,160 miles) in diameter: a figure rather more than one fourth that of the Earth. In comparison to its primary, the Moon is by far the largest satellite of the Solar system, and from that highly unusual situation, everything else follows.*

Let us suppose that Earth is exactly as it is, in size and rotation, but that only the stars are in the sky and that human beings can somehow exist and observe under such conditions.

Astronomy would then be a most dull and uninteresting science. The sky would seem to be a slowly rotating sphere, marked by tiny dots of light. There would be nothing to do with those stars but admire their beauty, observe their different brightness, colors, and patterns, and mark out arbitrary constellations. (And one thing more: From the stars alone, and from the difference in their positions with respect to the horizon as one moved about Earth's surface, one could argue, with considerable persuasiveness, that the Earth was a sphere.)

Add the Sun. We now have day and night, and from the way in which the stars change their position from night to night, it would be clear that the Sun, too, is moving about the Earth, but at a rate different from that of the stars. It could be persuasively argued that the Sun was embedded in a sphere as the stars were, but in a sphere that was transparent and which therefore could not be seen as it turned at its own characteristic speed different from that of the starry sphere.

Now add those planets that are visible to the naked eye—Mercury, Venus, Mars, Jupiter, and Saturn—so that the sky would now possess all the visible bodies but the Moon.

It would turn out that you must have a sphere for each of the different planets since each moves at a different rate. Furthermore, since the motions are not steady but vary in a rather complicated fashion, the working out of the rules governing the motions of those spheres would take a great deal of time, patience, and ingenuity—as, in fact, it did.

In the end, it would turn out that the structure was so unwieldy that it would become preferable to accept the less self-evident proposition that it was the Sun that was the center of the planetary system, not the Earth; and that it was the Earth that rotated in twenty-four hours, not the sky. This was the thesis finally advanced by the Polish astronomer, Nicholas Copernicus, in 1543.

It would seem, in short, that, without the Moon in the sky, the history of astronomy could develop in exactly the fashion that, in reality, it did.

* As of 1978, this is no longer true. Pluto seems to have a satellite nearly as large as itself.

—Except that we might well argue that it would not have done so without the Moon, for without the visible face of our satellite in the sky, it might be that mankind would never have had the impulse to study the sky in detail at all. He might merely have admired it.

What difference does the Moon make? Add it to the sky and see.

The Moon, like the Sun, Mercury, Venus, Mars, Jupiter, and Saturn, moves against the background of the stars, with its own characteristic speed, and requires a separate sphere of its own. It is for this reason that all seven are lumped together as planets "wanderers". (It is only in modern times that we have subtracted the Sun and Moon from the rest because of special characteristics in which they differ from the others.)

To be sure, the Moon moves more quickly than any of the other wandering bodies, but that in itself is not of great importance. It means only that the Moon is nearer the Earth than the rest are, and after all, one of the planets has to be nearest.

But of the planets, indeed of all the heavenly bodies, only the Sun, the Moon, and a very occasional comet ever appear as anything more than a point of light. Of these, comets appeared so rarely that they had no effect on ordinary human beings but to strike them with superstitious fear. The Sun, though an extended body, is too bright to be looked at for more than a moment, except when it is obscured by mist, and even then it appears merely a featureless (and therefore uninteresting) circle of light.

The Moon, on the other hand, is much softer in light and can be gazed at for indefinite periods of time. And the study, in however cursory a fashion, paid off at once, for, unlike the Sun, the Moon is not at all times a circle of light. The Moon changes its shape, moving through a regular cycle of phases. (It is not the existence of phases that is unique, but the fact that the Moon is close enough to appear large enough to make the phase-changes visible to the naked eye. Venus and Mercury also go through a cycle of phases as the Moon does, but they are too far away for the phases to be seen without telescopic help.)

The Moon's cycle of phases was ideally suited to attract attention. Because the Moon moves about the Earth in an only slightly elliptical orbit, it appears just about the same size at all times and the regularity of its cycle of phases is not confused by simultaneous changes in size and in speed of apparent motion across the sky. This made the phase-change a profitable field of study in the days when astronomy was utterly unsophisticated.

Furthermore, the Moon's phase-change goes through its complete cycle in a little over twenty-nine days, which is a particularly convenient length of time.

To the prehistoric farmer and hunter, the cycle of seasons (the year) was particularly important, but it was difficult to note that, on the average, the seasons repeated themselves every 365 and a fraction days. The number was too great to be kept track of easily.

To count twenty-nine or thirty days from each new Moon to the next, and then to count twelve or thirteen new Moons to each year, was much simpler, much more practical.

Therefore, once the regular change of phase of the Moon forced itself upon mankind, the marking of a calendar that would serve to keep track of the seasons of the year in terms of the phases of the Moon was the next step.

Alexander Marshak, in his book *The Roots of Civilization*, argues persuasively that, long before the beginning of history, early man was making stones in a code designed to keep track of the new Moons. Gerald Hawkins, in *Stonehenge Decoded*, argues just as persuasively that Stonehenge was a prehistoric observatory which was also designed to keep track of the new Moons, and to predict the lunar eclipses that came, occasionally, at the time of the full Moon (and are less frightening if one knows in advance they will come).

The fact that the cycle of the Moon's phases did not fit evenly into the rapid alternation of day and night, or into the slow alternation of the seasons (one synodic month=29.53 days=0.081 years) meant that, while the working out of a calendar was a simple enough task for primitive man to arrive at a useful approximation, there was sufficient complexity to it to force later generations into an ever-more sophisticated consideration of the comparative movements of the Sun and Moon.

It was the overriding practical necessity of working out a calendar based on the phases of the Moon, on the ever-changing shape of the face of the Moon, that forced human beings into astronomy. If the Moon were not in the sky, if the lunar calendar were not there to drive men into a careful study of the night sky, what else would do so?

Perhaps nothing, and perhaps we would be without astronomy today, and without the other branches of mathematics and science that the study of astronomy encouraged.

Then, too, the fact that the phase-change was so useful could not help but reinforce the notion of the existence of a benevolent deity, one who, out of his love of humanity, had arranged the skies into a calendar that would guide mankind into the proper ways of insuring a secure food supply. Each new Moon was celebrated as a religious festival in many early cultures, and the care of the calendar was usually placed in priestly hands. (The very word "calendar" is from the Latin word "to proclaim" since each month only began when the coming of the new Moon was officially proclaimed by the priests.)

We might conclude, then, that a considerable portion of the religious development of mankind, of the belief in God as a benevolent father rather than a capricious tyrant, can be traced back to the changing face of the Moon.

In addition, the fact that close study of the Moon was so important in controlling the daily lives of human beings could not help but give rise to the notion that the other planets might be important also. The face of the Moon may in this way have contributed to the growth of astrology and, through that, to other forms of mysticism.

To be, in large part, the foundation of some aspects of mankind's religion, mysticism, and science might seem enough responsibility to place on the fact of the existence in the sky of the face of the Moon, but we are not yet through.

The ancient Greek philosophers found it aesthetically satisfying to divide the Universe into two parts: the Earth and the heavenly bodies. To do so, they sought for fundamental differences in properties. Thus:

The heavenly bodies were all luminous; while Earth was nonluminous.

The heavenly bodies did not change; while on Earth everything grew and withered, rose and decayed, came into being and deteriorated.

The heavenly bodies moved in circular paths that were either regular or were irregular in a regularly-repeated fashion; while on Earth, the characteristic motions of objects was up and down or else totally irregular.

In short, the heavenly bodies were perfect; while Earth was not.

This sort of division of the Universe was elegant and symmetrical and had the virtue of pleasing academic minds, who tended to dismiss any evidence that would tend to disturb the pretty picture. And there *was* evidence against the picture despite the fact that it was held to by philosophers right down to 1600. In fact, there was a great flaw in it, and that existed in the face of the Moon.

It was clear to the unaided eye that the Moon, at least, of the heavenly bodies, was *not* luminous of itself but was as dull and as dark as the Earth. The relationship of the phases to the relative positions of the Moon and Sun made it clear even in ancient times that the Moon shone only by reflecting sunlight. The phase-change was not a real alteration in the shape of the Moon but was the result of the changing perspective from which a sunlit lunar hemisphere was viewed.

How could there be any doubt of this? Even if one disregarded the relative positions of Moon and Sun as representing a subtle argument convincing only to theoreticians, there remained the fact that when the light of the Sun was cut off by the intervening Earth, the brilliant gleam of the full Moon was slowly and progressively shaded. When the eclipse reached totality, the sunlight was cut off altogether (except for what little filtered past the Earth through its atmosphere) and the Moon was dark.

Another phenomenon that anyone could see came when the Moon was in its crescent phase and just a thin sliver of curving light. The rest of the Moon might then be seen sometimes shining with a dim ruddy light of its own. This was called "the old Moon in the new Moon's arms."

Was this dim, ruddy light a kind of feeble lunar luminosity only visible in the absence of sunlight? It could be more convincingly argued that this phenomenon demonstrated that the Earth, like the Moon, also shone by reflected sunlight. By the easily explained geometry of the situation, when the Moon appeared as a crescent to an observer on Earth, the Earth itself would appear as "full" to an observer on the Moon. The old Moon in the new Moon's arms was therefore lit by the splendid light of the full Earth, and we could see the dark of the Moon by its dim reflection of earthlight.

By observing the face of the Moon with the unaided eye, it was therefore possible to use the lunar phases, lunar eclipses, and the appearance at the time of the crescent, to demonstrate that the Moon and Earth were *both* nonluminous bodies and that *both* shone by reflected sunlight. The theoretical division was not as neat as it might have been. At least one heavenly body, the Moon, was very Earthlike in some ways, while the Earth was very heavenly-body-like in some ways.

Another flaw in the doctrine of the perfection of the heavens was revealed by the face of the Moon. That face was not an evenly shining object, as perfection would have required. There were smudges upon that face, most clearly and dramatically visible at the time of the full Moon. It was almost as though the Moon were dirty and decaying, almost as though it shared in the changeability that was thought to be characteristic of Earth and absent in the heavens.

All this might have been taken care of if philosophers had been willing to allow two worlds, Earth and Moon, each nonluminous and imperfect, while the remaining heavenly bodies could still be regarded as luminous and perfect.

This, however, was apparently unacceptable because too much in the way of philosophic and religious authority had been committed to asserting the uniqueness of Earth and its role as the only imperfect body and the only world.

The smudges on the face of the Moon and its nonluminosity could be explained, instead, by pointing out that of all the heavenly bodies, the Moon was closest to Earth and therefore most exposed to Earth's imperfections. It couldn't help catching a few. As for the question of earthlight illuminating the Moon that was ignored altogether till modern times.

Yet all the time that the philosophers constructed this neat picture of the Universe, folk wisdom hit closer to the mark (as it perhaps does oftener than scientists are willing to admit).

It was impossible for the average unsophisticated observer of the Moon not to attempt to make a picture out of the smudges upon its face. Given the natural anthropocentricity of mankind, it was very tempting to imagine those smudges to represent a man, and this is so in the best-known example in our own culture.

The "man in the Moon" was thought by some to be the one described in the Bible (Numbers 15:32–36) as having gathered sticks on the sabbath day. In the Bible he is said to have been stoned to death, but legends arose which described the man as having been placed in the Moon in further punishment. There, with him, is a thornbush, representing the sticks he was gathering, and a dog.

Thus, while the philosophers dismissed the smudges on the Moon's face as of no essential importance, laymen saw a man in the Moon. To the common people, the Moon was not only a world, it was an inhabited world.

In short, the face of the Moon, despite all that the more rigorous philosophers could say, gave rise to the concept of the plurality of worlds.

After all, once the notion arose that at least one of the heavenly lights might be more than a light, then it took but a small additional step to suppose that all the lights were more than lights. If the Moon was a world, and an inhabited world at that, then why not suppose all the heavenly bodies were inhabited worlds?

We don't know when the first tales of what we would now call space travel were told, but the oldest that survives dates back to the second century A.D. At that time, the Syrian writer Lucian of Samosata wrote of a ship that was carried up to the Moon by a waterspout. He describes the intelligent beings living on the Moon and tells of the war that they are conducting against the intelligent beings of the Sun over their conflicting ambitions to colonize the world of Venus.

The doctrine of the plurality of the worlds was not confined to legend makers and romancers, either. The deductions that could be made from the face of the Moon inspired heresy even within the ranks of the philosophers.

The German cardinal Nicholas of Cusa, in a book published in 1440, held that the Earth turned on its axis and moved about the Sun; that there was neither "up" nor "down" in space; that space was infinite; and, finally, that the stars were other suns and bore in their grasp other inhabited worlds in infinite numbers.

In all of this, he was very close to the views of modern astronomers, but Nicholas could advance no evidence for his views. However inspired, they were simply speculations, and they had no effect on the scientific establishment of his day.

For him, that was a good thing, for since his views created no stir, they roused no anger, and he was allowed to live out his life in peace.

But then, in Nicholas of Cusa's time, the establishment, both religious and philosophical, was peaceful and secure. A century and a half later, the Italian philosopher Giordano Bruno loudly proclaimed views similar to those of Nicholas at a time when religion in Europe had split into contending factions and when Copernicus's views were upsetting the established astronomical order. Dissent against orthodoxy could not be tolerated under such conditions and, in 1600, Bruno was burned at the stake for his heresies.

Even in Bruno's time, the arguments between those who upheld the doctrine of the plurality of worlds and those who accepted the uniqueness of Earth had to be largely a matter of words. There was no compelling evidence in either direction, only loud assertions.

But then, in 1608, the first primitive telescopes were constructed in the Netherlands and, in 1609, the Italian scientist Galileo Galilei built a somewhat better one for himself and turned it on the heavens.

Wherever Galileo looked, he made revolutionary discoveries. In looking at the Milky Way, for instance, he saw it to be made up of myriads of faint stars. And, indeed, wherever he looked he could see more stars through his instrument than he could with his unaided eye. This showed at once that there were objects in the sky that no man had ever seen before or could have seen before. For this, if for no other reason, it could be argued that the wisdom of the ancients had to be limited and must not be followed in blind reverence.

When studying Jupiter, Galileo found four small satellites circling it. That was visible proof that the Earth was not, after all, the center about which all heavenly objects revolved as the ancient concept of the Universe had it. At least four objects revolved about Jupiter, and the Copernican theory of planets revolving about the Sun seemed less absurd at once.

When studying Venus, Galileo found that it showed phases as the Moon did. Again, this was predicted by Copernicus's theory, but not by the orthodox view. It also presented visible proof that Venus, like the Moon and the Earth, was nonluminous and might, therefore, be a world. This was also borne out by the fact that, in the telescope, the various planets expanded into little moonlike circles of light. They, too, were extended bodies, and they looked like dots of light only because of their great distance.

Galileo even discovered that the Sun itself had black spots, a blow to the idea of the perfection of the heavens that struck at the very body which was viewed as the absolute ideal of perfection.

However, all these observations were at a rather lofty level. The satellites around Jupiter were small dots of light moving about a larger dot. The phases of Venus were just tiny crescents and half circles. They were not direct evidence of worlds, merely esoteric bits of data from which worlds might be deduced and, therefore, in themselves they might not have carried conviction.

But these observations did not stand alone. After all, given a telescope, what is the first object one would look at? Surely the Moon.

And that was what Galileo had done. Before he had looked at anything else, he had looked at the Moon. The face of the Moon, when studied by the unaided eye, had already given birth to the notion of the plurality of worlds. When it was expanded in the view of Galileo's telescope, it did more—for Galileo, quite plainly and simply, saw a world!

He saw mountain ranges and what looked like volcanic craters. He also saw dark, smooth patches which looked like seas. The Moon was not an unbroken and perfect silvery sphere, not even an unbroken and perfect dark sphere lit to silver by sunlight. It was very rough, very broken, very imperfect, very Earthlike. It was a world.

This was something there could be no mistake about. There were no deductions to be made, no chains of arguments to be carefully carried forward. It was eyewitness evidence.

After that, except for the case of astronomers so wedded to the ancient view that they preferred their textbooks to their eyes, there was no difficulty in accepting the concept of plurality of worlds. And once the Moon was accepted, all the other evidence fell into place.

The effect of Galileo's discovery of the worldness of the Moon is made plain by the fact that interplanetary romances suddenly came into popularity. The general public was clearly impressed.

The German astronomer Johannes Kepler wrote a story called *Somnium*, published posthumously in 1634. His hero was taken to the Moon by spirits, and though Kepler's Moon had no intelligent beings on it, it bore strange animals and plants that grew rapidly during the two weeks of the lunar day and then died.

An English clergyman, Francis Godwin, wrote a much more popular story called *Man in the Moone*, published in 1638. Godwin's hero flew to the Moon in a chariot hitched to great geese which were supposed to migrate to the Moon regularly. Godwin described the Moon as a world much like the Earth, but better.

In 1650, Cyrano de Bergerac, the French artist and duelist, published *Voyage to the Moon*, in which he worked out seven different fanciful ways of making the trip. One of them involved the use of rockets which, as it happened, was the correct way. It is interesting that Cyrano suggested this method in a science-fiction story, a full generation before the scientific theory behind it was firmly established by the English scientist, Isaac Newton.

The tide of interplanetary romance that had been inspired by Galileo's view of the expanded face of the Moon never completely died down. Edgar Allan Poe, Jules Verne, and H. G. Wells all wrote stories about trips to the Moon, and after Wells, space travel became the staple of science fiction generally.

And, again, it was the existence of the face of the Moon, which had already played such an important role in the development of human thought, that made it possible for interplanetary travel to be translated from romance to fact.

That the Moon appears as large as it does in the sky—that there is a visible face to the Moon—comes about because of a combination of its size and its closeness to us.

It is 381,000 kilometers (237,000 miles) from Earth. This is a great distance but certainly not a *very* great distance. It is only 9.5 times as great as the circumference of the Earth, and the first primitive rockets capable of carrying man into space could cross that distance in only three days. Allow for a day of exploration and then the return, and the first attempt at exploration of the Moon would take no more than a week altogether.

It is a challenge but not an impossible challenge.

Imagine what the situation would have been if there were no face of the Moon in the sky, if the Moon didn't exist. In that case, the nearest sizable body to ourselves would be Venus which, even at its closest, is fully 40,000,000 kilometers

(25,000,000 miles) from Earth, or 105 times as far away as the Moon is.

Venus couldn't possibly be explored in a round trip of a week. A year and a quarter is more like it. Faced with the task of remaining pent up in a spaceship for a year and a quarter in order to make the very simplest possible interplanetary exploration, it seems doubtful that anyone would ever have dreamed of trying it.

The Moon is necessary as a first step we can take, as a child's game on which we can test ourselves and develop to the point of taking on the real thing.

Or suppose that the Moon did exist but that it didn't exhibit an actual face, either because it was farther away than it is or smaller than it is. If it were 3,200,000 kilometers (2,000,000 miles) away, it would appear in our sky as just a bright dot of light, and would not show a clearly distinguishable face. In that case, we might count on a round trip of two months rather than a week. Perhaps a little too rough for a first step.

And if it were small? The smaller the Moon, the less worldlike it would be and the less interest it would draw. It might not represent enough of a challenge to force us into doing more than sending an unmanned probe in its direction.

In any of the cases, then, in which our sky lacked the face of the Moon, it might be that we would develop the space age to the point of setting up useful satellites of all sorts in Earth's immediate vicinity, and that we might even send out unmanned probes to explore the bodies of the Solar system, but no more than that. It might never occur to us to send out manned ships since there would be no interesting object in space that we could dream of reaching, or no object we could reach that would be of more than minimal interest.

In short, it is the fact that the face of the Moon exists in the sky that has made the development of space travel possible. It is the face of the Moon, through a series of influences stretching far back into the mists of prehistory, that made it possible for a spaceship to land on the Moon on July 20, 1969, and for two men to emerge and step out upon—the face of the Moon.

Unlike most of the essays in the book, this one was not written for a magazine. A friend of mine, Tom Purdom, conceived the notion of a book containing the stories of specific scientific discoveries as told by professional writers with solid knowledge of science. Naturally, he found the writers he needed among the list of those who wrote for the science-fiction magazines, for it was there that one was most likely to encounter the double skill needed.

For the book, I wrote the following essay. Since the book is now, I believe, out of print, I might as well rescue the essay and place it here.

11

THE DISCOVERY OF ARGON

One of the major excitements in scientific research is that every once in a while, when you go out rabbit hunting, you bag a bear.

And John William Strutt, the English physicist (better known as Lord Rayleigh, a title he inherited in 1873, when he was thirty-one), bagged a large bear, very much to his own surprise.

The key to Lord Rayleigh, and to his great adventure, was his clean mathematical mind, his absolute obsession with that final decimal point, that last bit of neatness. At Cambridge, he attained the title of Senior Wrangler—top man in his class in mathematics. And when he wrote a paper, he wrote it with such precision it could be sent to the publishers as written, without revision.

Rayleigh was a physicist, who ranged through almost every branch of the subject. He wrote a treatise on sound in 1877 that reduced the phenomenon to a branch of mechanics—matter in motion. He worked out in satisfactory detail the manner in which light is scattered in the atmosphere, explaining, in this way, why the sky is blue. The phenomenon is still called "Rayleigh scattering."

Even when his analysis did *not* fit the facts, it turned out to be important. An equation he derived for the details of heat radiation did not explain what was actually observed to happen. So clear and irrefutable a piece of logic and mathematics was it, however, that the nonfit was clearly a matter of first importance. Rayleigh founded his analysis on physical basics and, therefore, those basics had to be wrong. His equation killed "classical physics," and what we now call "modern physics" had to be put in its place by the united labors of a generation of physicists, to the great improvement of our understanding of the Universe.

It was this meticulous mind, then, that in 1882 turned to the question of the structure of the atoms of the elements.

Chemistry was not Rayleigh's forte; he did not like it. One important chemical problem, however, appealed to the formidably precise instrument that was his mind.

Scientists, generally, are strongly drawn to the thesis that there is an underlying simplicity in nature. Where there seems to be disorder and jumble, the search grows intense for a possible basic order beneath, a possible relationship that will pull the apparent mess into a neat design.

In 1882, there were already known some seventy different chemical elements, each with a different atomic weight. There seemed no logic to these atomic weights. The atoms of each element possessed a mass that bore no particular relationship to that of the atoms of any other element.

Yet, when the atomic weights had first begun to be determined eighty years before, some order had seemed to appear. The atom of oxygen had, it seemed, just sixteen times the mass of an atom of hydrogen; the nitrogen atom had just fourteen times the mass of a hydrogen atom; the carbon atom had just twelve times the mass.

In 1815, an English physicist, William Prout, had suggested that this was because all the atoms of elements other than hydrogen were built up of hydrogen atoms. The oxygen atom was sixteen times the mass of a hydrogen atom because it was built up of sixteen hydrogen atoms and so on. This would reduce all the elements to different combinations of a single atom and greatly increase the simplicity of the Universe.

Unfortunately for "Prout's hypothesis," it became apparent, as more and more atomic weights were determined, that a number of elements had atoms with masses that were *not* even multiples of hydrogen. It was rather disappointing that this should be so, and chemist after chemist returned rather wistfully to the problem. Perhaps if atomic weights were determined more carefully (and the determination was a tricky business indeed), some order might be wrenched out of them.

Unfortunately, no. The more painstaking the determinations, the less order and the more jumble there seemed to exist.

Yet a great many of the elements *did* have atomic weights that seemed exact multiples of that of hydrogen. If the atoms were not built out of hydrogen, why were there so many exact multiples? It was asking too much of coincidence.

But were even those atoms with atomic weights that were multiples of hydrogen, really *exact* multiples? Perhaps they were only *approximate* multiples. Some determinations seemed to show that it was a matter of approximation rather than exactness.

It was important to know this. If atomic weights were not exact multiples, then it was actually harmful to suppose they were. Anything that gave reason to cling to attractive, but false, views hurt science by delaying research in other, and more truly, fruitful directions.

Here was where Rayleigh came in. What was needed were precise measurements—measurements more precise than any that had yet been made, measurements that would leave all doubt behind and settle the matter once and for all.

Rayleigh decided to concentrate on a very few elements and go all out for precision. He began with the classic pair, hydrogen and oxygen. From the very earliest days of atomic weight determinations, it had seemed that the oxygen atom had a mass that was just about sixteen times that of the hydrogen atom. Well, was it *exactly* sixteen times?

Oxygen and hydrogen are both gases made up of molecules, which are, in turn, made up of pairs of atoms. This means, according to chemical theory, that if both are perfect gases, their densities will be in exact proportion to their atomic weights. If Rayleigh were but to measure the densities of oxygen and hydrogen with great precision, he would have what he wanted.

Except that hydrogen and oxygen are not quite perfect gases; almost, but not quite. The more rarefied they are made, however, the smaller the pressure exerted by them and upon them, the more nearly they approach the perfect. They become actually perfect when pressure reaches zero, but at zero pressure, meaningful density measurements cannot be made.

What Rayleigh did, then, was to measure the density under different pressures, and observe how that density changed as the pressure decreased. From this, he could calculate what it would be at ordinary pressure, if the gas were perfect.

This can be briefly said, but Rayleigh's meticulous measurements under different conditions, his checks and rechecks, his laborious attempt to account for every possible source of error, and finally his use of oxygen and hydrogen obtained from a variety of sources by a variety of chemical procedures, took ten years.

By 1892, he was able to say that the atomic weight of oxygen was *not* exactly sixteen times that of hydrogen. It was 15.882 times.

Prout's hypothesis did not hold. Where an atom seemed to have a mass that was an exact multiple of hydrogen, it didn't really. At least it was true of oxygen and perhaps it was true of any element. Even as the final decision on oxygen was approaching, Rayleigh had begun work on nitrogen.

Nitrogen, like oxygen, is a gas; it, too, is made up of molecules that are in turn made up of a pair of atoms; it, too, while not a perfect gas, is fairly close to being one and gets closer with decreasing pressure. In short, the procedures Rayleigh had worked out for hydrogen and oxygen should also work for nitrogen, and should get results in far less than ten years.

The obvious source of nitrogen was the air, which is a mixture composed of nitrogen and oxygen in the ratio of 4 to 1, together with the admixture of small quantities of other materials: water vapor, carbon dioxide, smoke, dust, and so on.

It is easy to obtain the nitrogen since of all the components of air, it was the only one known to remain unchanged under ordinary physical and chemical attack. Air can be filtered clean of dust, for instance; it can be reduced to low temperatures to freeze out the water; it can be subjected to chemical treatment to remove carbon dioxide and oxygen. Nitrogen, being "inert," withstands and survives all this. In the end, then, what was air is altered to nitrogen, and it was this nitrogen that Rayleigh used for his density determinations.

The density of nitrogen, casually determined, was taken as fourteen times that of hydrogen. Rayleigh's careful measurements showed it was 13.97 times, close to the exact multiple, but not quite. Rayleigh was sufficiently confident of his result to feel that it would not be justified to "round it off" to 14.00.

Still, he knew he would feel better if he prepared nitrogen in another fashion and checked the density again. He adjusted his procedure in such a way that some of the nitrogen he obtained would come from a nitrogen-containing compound called "ammonia."

And here came a blow!

The density figure he now got was very slightly less than that he had obtained for nitrogen from air alone. Was something wrong with his ammonia? He altered his procedures further to get more and more of the nitrogen from ammonia and the discrepancy grew worse. Finally, he worked with nitrogen obtained *entirely* from ammonia and his density figure was then 13.90, instead of the 13.97 he had found for nitrogen from air.

To another, that small discrepancy, amounting to only one half of 1 per cent, might hardly have seemed important. After all, neither 13.90 nor 13.97 represented an exact multiple, so Prout's hypothesis once more had been disproved. Why worry further about so small a discrepancy?

To Rayleigh, however, this discrepancy was insupportable. That it existed at all was an insult to his precision. Besides, it simply *shouldn't* be there. He had, after all, duplicated his figures for hydrogen and oxygen from a variety of sources. Why not with nitrogen?

Rayleigh abandoned all thought of Prout's hypothesis and turned his intellect to this new and thoroughly unexpected problem. The discrepancy, small as it was, acted upon him like an itch he could not forbear scratching.

Rayleigh might have assumed there were two slightly different kinds of nitrogen, one of which occurred in air and one in the compound, ammonia. This, however, in view of all scientific findings in the eighteenth and nineteenth centuries, seemed in the highest degree unlikely.

What was much more likely was that one (or both) of his samples of nitrogen had a small but consistent impurity present—an impurity whose presence altered the density figures.

For instance, a molecule of ammonia is made up of one atom of nitrogen and three atoms of hydrogen. Suppose that in addition to the nitrogen obtained from the ammonia, a small amount of the hydrogen was carried along. The hydrogen is much less dense than nitrogen and even a small quantity of hydrogen—about 0.5 per cent—mixed with the nitrogen would suffice to lower the overall density of the presumably pure nitrogen and make it sufficiently less than the density of the nitrogen from air (which contains no hydrogen to speak of) to account for the discrepancy.

The trouble with that notion was that hydrogen is considerably different from nitrogen in its properties and even 0.5 per cent of hydrogen in nitrogen would have been easy to detect. The meticulous Rayleigh could detect none and was forced to conclude that no significant amount of hydrogen was present.

There was a second possibility, though. The nitrogen molecules consist of two nitrogen atoms apiece, the two atoms being so closely associated that the molecule can be considered a single particle. But suppose some of the nitrogen molecules broke apart and liberated single nitrogen atoms?

The single-atom nitrogen would have only half the density of the two-atom nitrogen molecules. If 1 per cent of the molecules split up, the density of the nitrogen from ammonia would be lowered by the observed amount, compared with the density of the nitrogen from air.

But there were a couple of catches here. It seemed a well-established chemical fact that nitrogen molecules did not split up into single atoms under the conditions which were used to isolate nitrogen from ammonia. And, if somehow the nitrogen molecules did split and managed to stay split after having been prepared from ammonia, why did not the molecules in nitrogen from air split up when subjected to the same conditions? They did not. Their density stayed high.

The single-atom nitrogen idea had to be eliminated and there seemed nothing else that could affect the density of the nitrogen from ammonia.

Rayleigh eventually cleared this variety of nitrogen even further. He prepared nitrogen from nitrogen-containing compounds other than ammonia. This means other types of chemical reactions were used and other possible impurities were encountered.

The densities he obtained for nitrogen procured from any compound corresponded with what he found in nitrogen from ammonia. Could it be that with a variety of origins and of preparations, all the nitrogen samples obtained from various compounds somehow ended with the same quantity of the same indetectable impurity that lowered its density by the same amount? Or with quantities of a variety of indetectable impurities that always, somehow, managed to lower its density by the same amount?

No! That was far too much to accept.

It was easier to believe that each of the nitrogen samples obtained from different compounds was pure nitrogen and for that reason they all showed the same density figure.

Standing out against the unanimous situation among the nitrogen samples from the different compounds was the one and only exception: nitrogen from air. There had to be something wrong with the nitrogen from the air. It was too dense; it must, therefore, contain some impurity that was denser than nitrogen itself. What is more, it had to be some systematic impurity that wasn't local or temporary because Rayleigh always got the same too-high figure from any sample of nitrogen he obtained from the air.

Rayleigh had obtained the nitrogen from air by the common method generally used by chemists. He had subtracted everything else, relying on the inertness of nitrogen to keep it untouched, while chemicals combined with and washed out carbon dioxide, oxygen and, presumably, everything but nitrogen.

But was that so? That was just an assumption and perhaps an unjustified one. Could some carbon dioxide, or oxygen, or both have been left behind by a chemical broom that didn't quite sweep clean? Both carbon dioxide and oxygen were denser than nitrogen and any, if left behind, would raise the density of nitrogen from air above the figure for pure nitrogen.

Take carbon dioxide first. It is one and one-half times as dense as nitrogen, and if the nitrogen from air were only 0.5 per cent carbon dioxide that would account for the discrepancy. However, there is very little carbon dioxide in the air; it makes up only 0.03 per cent of the whole. Even if all the carbon dioxide had been left behind, it would have been far from enough to account for the discrepancy.

What about oxygen? Well, it is only about one and one-seventh times as dense as nitrogen. The final nitrogen obtained would have to be 4 per cent oxygen to be sufficiently high in density to account for the discrepancy.

And that much oxygen was impossible. If the nitrogen from air, with which Rayleigh was working, had been 4 per cent oxygen in reality, that quantity of oxygen should have been easily detectable and, what is more, easily removable. If that much oxygen were added to nitrogen from ammonia, for instance, in order to wipe out the discrepancy, that oxygen could be easily removed and the discrepancy restored in full.

So the nitrogen from air did not have its density raised by appreciable quantities of the obvious impurities.

What about unobvious impurities? Oxygen consists of molecules made up of two atoms apiece, but there are ways in which oxygen can be forced into molecules of three atoms apiece. An electrical discharge through oxygen, for instance, would form small quantities of such three-atom molecules, which are called "ozone." Ozone, with its three-atom molecules, is one and one-half times as dense as ordinary oxygen with its two-atom molecules.

Could it be, then, that nitrogen, too, could form three-atom molecules? Let us make up the name "nitrozone" for such a three-atom nitrogen molecule. Such nitrozone would then be one and one-half times as dense as ordinary two-atom nitrogen. If 1 per cent of the supposed nitrogen from air were really nitrozone, that would account for the discrepancy.

But wait. Ozone can be formed from oxygen only with difficulty and, left to itself, it quickly breaks down into ordinary oxygen again. The same conditions which force oxygen to form ozone, do not force nitrogen to form nitrozone. One can only assume that nitrozone is harder to form than ozone and breaks down more easily when it does form.

Why, then, should nitrozone form and remain in nitrogen from air, when it never forms and remains in nitrogen from some chemical?

The only way of making sense out of this dilemma is to suppose that nitrozone does not form at all.

Rayleigh had come to a dead end. It seemed quite clear there was something wrong with the nitrogen he obtained from air, that its density was too high, and therefore, that it contained some impurity that was denser than nitrogen itself, probably considerably denser.

What stopped him was that all possible impurities seemed to have been eliminated, and if there was no impurity, then the discrepancy in densities could not be explained.

Rayleigh took the rather unusual measure of appealing for help. On September 24, 1892, he threw the problem open to the scientific community, writing a letter to the scientific journal, *Nature*, outlining the problem and asking for suggestions.

Nature ran the letter in its issue of September 29, 1892, but for over a year, it produced not a single response. The plea to the scientific community drew a blank, at least for a while.

Rayleigh continued to work on his own. He continued to play with the notion of nitrozone; not that it seemed in the least likely, but because all the other alternatives he had thought of seemed far less likely still.

The letter was seen, however, by a Scottish chemist, William Ramsay, ten years younger than Rayleigh, and Ramsay brooded on the matter for some time.

Ramsay was an all-around man. As a youngster, he was interested in music and languages and then developed further interests in mathematics, and science. He was also a man of athletic inclinations and was, for instance, a spectacular diver. Whatever he turned his hand to he did well, and when he established a home chemistry laboratory while still a youngster, he taught himself to blow glass like an expert. To the end of his life, he made most of his glass equipment himself.

While Rayleigh was struggling with gas densities, Ramsay was a professor of chemistry at University College in London. The younger man was primarily interested in organic chemistry (the chemistry of carbon compounds), and matters involving pure nitrogen gas and density measurements would not ordinarily have been expected to interest him. But they did.

Ramsay approached it with a chemist's cast of mind, and he was less easily satisfied with the basic method for obtaining nitrogen from air than Rayleigh had been. It is perhaps reasonable to suspect that Rayleigh accepted the subtraction method of removing everything else from air because that was standard chemical operating procedure. Who was he, a physicist, to question chemists in their own field? Ramsay, a chemist himself, was perfectly ready to question them.

Ramsay reasoned that to obtain nitrogen by removing everything else was dangerous. It was a process that assumed that nitrogen and *only* nitrogen would remain untouched. It ended by defining nitrogen through a series of negatives.

Nitrogen did *not* filter out as dust did; it did *not* freeze out as water vapor did; it did *not* react with a base as carbon dioxide did; it did *not* react with a reducing substance as oxygen did.

But what if along with nitrogen there was some other gas (or ten other gases, for that matter) which also did not filter out, or freeze out, or react with a base, or react with a reducing substance? It would remain with the nitrogen and be called nitrogen by virtue of its agreement on negatives.

No, no, surely one had to find something that nitrogen *did* do and use that positive property to distinguish it from other substances—as in the case of every other element.

Ramsay had another advantage over Rayleigh (and over most other chemists as well, alas). He knew the history of his own science. He knew that a century earlier, in 1785, an English chemist, Henry Cavendish, had also had this notion about trapping nitrogen through positives rather than negatives.

It had been very early in the history of the chemical investigation of air, but Cavendish was a supremely great experimenter who was a century ahead of his time in a number of different directions. He found that nitrogen was not entirely inert. It could indeed be made to combine with other elements under extreme conditions.

For instance, if an electric discharge was passed through air, some of the nitrogen and oxygen in the air would combine. Cavendish therefore determined to remove the nitrogen from the air by forcing it into combination with oxygen in this manner and then dissolving away the nitrogen-oxygen compound formed. Then he could determine if there were anything in air that did *not* behave in this manner and, was therefore, *not* nitrogen.

When electric discharge caused no further shrinkage in the volume of air with which Cavendish was working, it might have been that all the nitrogen was gone and that what was left was another gas that would not combine with oxygen. In that case, adding more oxygen would introduce no further change.

On the other hand, it might be that all the oxygen was gone and what was left was excess nitrogen. In that case, if more oxygen were added, the volume would continue to shrink.

Cavendish added more oxygen and the volume continued to shrink. It shrank until his original sample of air was reduced to a bubble which he estimate to represent $^1/_{120}$ of the original quantity. That final bubble stayed. Though he added more oxygen in quantity, no change was induced. When the oxygen was removed, there was that bubble.

The conclusion seemed clear, at least to Cavendish's remorselessly logical mind. He maintained that there was an additional gas in air, present in small quantities—less than 1 per cent of the total. This additional gas was even *more* inert than nitrogen and would not combine with other elements even under conditions where nitrogen would. The additional gas therefore was different from nitrogen.

Cavendish announced all this but, unfortunately, he was an eccentric man, who avoided all human company to an almost insane degree. He was completely indifferent to fame and much of his greatest work he left unpublished altogether, to remain undiscovered until years after his death. What he did publish, he did not particularly publicize. If people refused to believe him or listen to him, he was supremely uncaring.

So Cavendish's discovery of a new gas in air went by the board, and his work was forgotten except by a few—like Ramsay.

Ramsay thought of Cavendish's experiment in connection with Rayleigh's problem. What if nitrogen from air contained a small quantity of this additional gas Cavendish had spoken of and what if that additional gas were considerably more dense than nitrogen?

When nitrogen from the air was forced (by the electric discharge of lightning, or by any other phenomenon) to combine with oxygen or any other element, this other gas, more inert than nitrogen, would remain untouched. Nitrogen compounds would form, but not compounds of this other gas.

When nitrogen was formed from compounds, therefore, it would appear pure, uncontaminated by the other gas. When nitrogen was prepared from the atmosphere, it would be contaminated. In that case, the nitrogen (or supposed nitrogen) from air would naturally be a little denser than the pure nitrogen from compounds.

Mere reasoning, however, was not enough. The presence of such a gaseous impurity of greater density than nitrogen itself would have to be demonstrated. In a fever of excitement, Ramsay wrote to Rayleigh in early 1894 and asked permission to tackle the problem along these lines. Rayleigh, rather relieved to have a first-class chemist on his side, gave this permission gladly and promptly.

Ramsay did not wish to follow Cavendish's method exactly. Adding oxygen to the nitrogen would always raise problems by introducing a gaseous impurity. There might be a question as to whether all the added oxygen had been removed again.

Rather, he chose to use a solid, the very active metal, magnesium. Magnesium, when heated, combines with oxygen so rapidly that it bursts into white-hot flame. So active is it that, in default of something like oxygen, it will even snatch at the usually inert nitrogen. If magnesium is raised to red heat in a pure nitrogen atmosphere, it combines with nitrogen to form a yellow solid, magnesium nitride. In this way, Ramsay neither would use nor form a gas.

Preparing nitrogen from air in the usual manner, Ramsay then passed it back and forth over red-hot magnesium and watched the nitride form. He waited excitedly. If, indeed, there was a small quantity of another gas present and if, indeed, it were even more inert than nitrogen, then it might not combine even with red-hot magnesium.

The quantity of nitrogen grew less and less till it reached a mere bubble about one-eightieth the size of the original gas volume. And that was it!

This final bubble did *not* react with magnesium, or with anything Ramsay tried. It was completely inert, much more inert than nitrogen and it was, therefore, a gas that must be distinct from nitrogen. He collected enough of it to test its density and found it was just about one and one-half times as dense as nitrogen. The quantity and density of this gas was just enough to account for the discrepancy that Rayleigh had puzzled over for at least three years.

When Rayleigh was informed of these results, he was most cautious. To suggest a new element to explain the discrepancy was uncomfortably like a *deus ex machina*. He preferred an explanation based on what was already known—such as the substance we have been calling nitrozone, the three-atom nitrogen molecule. After all nitrozone would be expected to have a density just one and one-half times that of nitrogen.

Ramsay, the chemist, could not accept nitrozone quite so easily. Rayleigh, the physicist, might concentrate on a physical property such as density, but Ramsay, the chemist, knew that by all the laws of chemistry, nitrozone had to be extremely active. It couldn't possibly be as inert as this new gas seemed to be.

Ramsay, the chemist, therefore searched for some additional method of distinguishing between a new element, and something that was merely a new form of an old element.

As it happened, a generation earlier, the technique of spectroscopy had been developed. By this technique, an unknown substance could be heated till it glowed with a light developed within itself. This light, passed through a spectroscope was broken into the separate colors that made it up, and there appeared as a pattern of lines. Each element produces a line pattern that is unique unto itself and that, in effect, acts as a "fingerprint" for that element.

The new gas was heated, therefore, till it glowed, and the light was passed through a spectroscope. Faint lines, associated with nitrogen, were indeed present, which showed that not all the nitrogen had been removed by the hot magnesium; enough remained to be detected by the delicate technique of spectroscopy. In addition, though, there were several lines of red and green which formed a pattern never before seen in connection with any known element.

That was conclusive! All thought of nitrozone had to be abandoned and Rayleigh admitted that what Ramsay had uncovered was a new element. In August 1894, these results were announced before a meeting of British chemists.

The chairman of the meeting, hearing the description of this new and unprecedentedly inert gas, suggested the name "argon" for it, from a Greek word, meaning "inert."

The suggestion was accepted and thus, argon, discovered through a line of research that had been aimed in an entirely different direction, entered the family of elements.

Once that was done, Rayleigh (with perhaps a sense of relief) returned to his beloved physics and left the further investigation of the matter to Ramsay. Ramsay went on to search for other gases like argon and in the course of the next four years discovered four more—helium, neon, krypton, and xenon—which were all much rarer than argon.

Soon after argon was discovered, the phenomenon of radioactivity was detected and blazed like a comet across the skies of science. In the light of radioactivity and all that followed, physicists and chemists came to realize that atoms had a complex internal structure. As it happened, the properties of the newly discovered argon and its sister gases proved to be of special help in determining the nature of certain aspects of this structure.

The new view of atomic structure eventually showed that Prout's hypothesis was more nearly right than wrong after all. It had seemed wrong only because there were, indeed, different varieties of the elements; varieties so subtly different that the techniques of the nineteenth century did not suffice to demonstrate their existence.

Rayleigh's meticulous determination of densities, in other words, was interesting but, as it turned out, not really a matter of first importance. Atomic weights were *not*, after all, fundamental to the structure of elements.

Out of Rayleigh's work on densities, then (which neither he nor any other scientist realized was basically unimportant to atomic theory), there arose a small discrepancy which resulted in a completely unexpected discovery, which turned out (as neither he nor any other scientist could have foreseen) to have great importance to atomic theory.

That is how science works sometimes.

Nor did Rayleigh and Ramsay have to wait long for full recognition. In 1904, each received a Nobel Prize—a *different* Nobel Prize. Rayleigh received a Nobel Prize in physics for his density determinations, and Ramsay received a Nobel Prize in chemistry for his detection of argon.

Water, in itself, is not living, but it certainly belongs in a book that deals primarily with life.

Life developed in water, still lives in water, must have water at every moment as the background of the chemical changes that are life.

You would think that man would be intelligent enough not to outbreed so vital a resource and certainly not to foul the resource as he is in the process of outbreeding it. Well, if you thought that, you would be wrong.

12

WATER

Do you know how large a cubic kilometer is?

Imagine a hollow cube—one kilometer long, one kilometer wide, one kilometer high—and imagine that cube full of water. Imagine that cubic kilometer of water emptied upon the island of Manhattan. If that water remained on a Manhattan imagined to be bare, level land and did not run off, it would cover the entire island to a depth of 17.5 meters. The depth of that water would be equal to the height of a five-story building.

That is *one* cubic kilometer of water.

The total amount of water on Earth is 1,280,000,000 (one and a quarter *billion*) cubic kilometers.

All that water is pulled by Earth's gravitation to the lowest portions of Earth's surface and there it makes a puddle called the ocean that covers 70 per cent of that surface. Only 30 per cent of Earth's surface is high enough to emerge above the top of that puddle to form the continents and islands.

The ocean has a surface area of 360,000,000 square kilometers, over sixteen times that of the Soviet Union. In at least one spot, it is just over 11 kilometers deep and, on the average, it is 3.7 kilometers deep.

What's more, this water is the permanent possession of our planet. There is a thin drizzle of water molecules that, through complicated processes, is lost to outer space, but it would take many hundreds of millions of years for that loss to become perceptible. Some water is consumed in various geological, chemical, and biological changes, sometimes with and sometimes without human intervention and encouragement. Always, however, the process comes full circle and the water is produced again. One and a quarter billion cubic kilometers of water is what we have, did have, and will have.

How, then, can we speak of a water shortage?

If the water of the ocean were pure water, we couldn't—but it isn't pure water. Water is an excellent solvent and the ocean contains not only water but a variety of solid materials, with four-fifths of that solid material consisting of sodium chloride, which is ordinary table salt. Every cubic kilometer of ocean water contains within it some 40,000,000,000 kilograms of dissolved solids.

The ocean contains myriads of forms of life that are adapted to, and flourish in, such a salt solution. Land life, however, whether plant or animal, cannot make do with ocean water. If human beings had at their disposal ocean water *only*, there would be no water they could drink,* no water with which they could wash, no water with which they could irrigate their crops, no water which they could use for their industrial processes. There could be, in short, no human civilization and, indeed, no human life.

If gravity were the only force to which Earth's water was subjected, all of it would be in the ocean and all of it would be salt. Earth's land surface would be bone dry and, except along the shores, as sterile as the Moon.

But this is not so. There is one source of energy that can suffice to lift water out of the ocean on a large scale, and that is the Sun's heat. The ocean's surface evaporates, particularly in the warm tropical regions, and vast quantities of water vapor permeate the atmosphere.

The atmosphere can only hold so much water vapor, however, and cold air can hold less than warm air can. When the water vapor lifts into the cold upper regions of the atmosphere

* "Water, water, everywhere," said the Ancient Mariner, becalmed in the Pacific, "nor any drop to drink."

or when it drifts north or south, away from the tropics and into cooler regions, that water vapor condenses into clouds made up of fine droplets of water or crystals of ice. Eventually, the water precipitates out of the atmosphere as either rain or snow and is restored to the ocean from which it came.

Only about 1/30,000 of the Earth's water supply is in the atmosphere as water vapor at any one time, but this is the equivalent of 45,000 cubic kilometers of water and is a sizable amount in human terms.

The crucial point is this: When ocean water evaporates, only the *water* turns to vapor; the solid matter dissolves in the ocean and remains behind. This means that the water vapor in the atmosphere, and the rain or snow that forms from it, is "fresh water" and it is fresh water which human beings can use for drinking, washing, agriculture, and industry.

But where does the fresh water go when it precipitates?

Most of it falls directly into the ocean, of course, and is at once mixed with and lost in the salt water. Some of the masses of water vapor drift over the land surface of the Earth, however, so that rain and snow fall there, too.

Eventually, of course, gravitation is paramount and any fresh water falling on Earth's land surface returns to the ocean. This, however, takes time. At any one given moment there are about 33,700,000 cubic kilometers of fresh water in the atmosphere and on land. This means that of all the water on Earth, 97.4 per cent is ocean water and 2.6 per cent is fresh water.

Even a mere 33,700,000 cubic kilometers of fresh water sounds as though no one can ever imagine human beings using it all and needing more, but let us go on.

Fresh water can fall upon Earth's land surface as either rain or snow, depending on the temperature of air and land. If it falls as rain, it will sink through porous soil and rock until it finally and inevitably reaches a nonporous rock layer. It then piles up there as groundwater.

Gravitational pull will cause the groundwater to percolate to lower ground and eventually reach the ocean again. En route, it can encounter ground low enough for it to emerge on the surface and collect in ponds and lakes or in springs and rivers. Almost all this open water returns to the ocean more rapidly in free flow than it would by percolating through the soil. (Some of the water, of course, evaporates directly and may then be restored to the ocean through precipitation, or may precipitate back on land to go through the cycle again.)

And as liquid water on land pours or percolates back into the ocean, some 120,000 cubic kilometers of rain or snow falls on the land surface each year to renew the supply.

If it is snow that falls on the land surface, it tends to accumulate, for it is solid and does not flow in the usual sense. In the warmer seasons, however, the snow may melt and then it would behave as though it had fallen as rain.

In very cold regions, there may not be enough summer warmth to melt the winter snow entirely, and in that case, snow accumulates and piles up from year to year and, under the pressure of its own weight, becomes ice. In Antarctica, for instance, there is an ice sheet covering over 14,000,000 square kilometers of land (one-half times the area of China or the United States) to an average depth of a little over two kilometers. The volume of accumulated ice in Antarctica thus amounts to about 30,000,000 cubic kilometers. A far smaller quantity exists in Greenland, and there is a scattering in other polar regions and on the mountain heights.

All told, there are 33,000,000 cubic kilometers of ice on Earth and this represents just about 98 percent of all the fresh water on Earth.

Ice does not continue to accumulate endlessly, of course. Its own weight causes it to flatten and push outward. At the ocean rim of the ice sheets, great pieces break off and float in the ocean as "ice islands" and "icebergs," until they melt and are restored to the ocean. From the mountaintops, rivers of ice called "glaciers" are forced to lower levels where they melt.

Ice, although a form of fresh water, is, on the whole, unavailable for use. Land surface covered with a permanent ice layer is as sterile as land with no water at all. Antarctica is the most life-free portion of our planet.

That leaves us with only liquid fresh water available for direct use by land life, and the total quantity of liquid fresh water on Earth is only 645,000 cubic kilometers.

This is only $^1/_{20}$ of 1 per cent of all the water on Earth, and it is with that $^1/_{20}$ of 1 percent that we must make do.

Can 645,000 cubic kilometers of usable water be sufficient? If we imagined such a supply to be evenly divided among the human population of the Earth, that would amount to 160,000 cubic meters for every human being on Earth. What's more, if each human being had an equal share of the rain or snow that fell, he would get 30,000 cubic meters of new fresh water each year to replace that which he used or wasted.

How does this compare with human needs?

Suppose we consider the United States, where water is used most lavishly on a per capita basis and where, all things being equal, the pinch might be felt first.

Assuming an average American will drink eight glasses of water a day, he will consume 0.6 cubic meters in a year, which is virtually nothing. Water is used for other reasons about the home, though, for everything from washing the dishes to watering the lawn to bathing. All told, the average American consumes at home 200 cubic meters of water per year.

Even that is not very much. We need a great deal of water for our domestic animals, for our growing crops, for our industries. As an example, to make a kilogram of steel requires 200 kilograms of water, and to grow a kilogram of wheat requires 8,000 kilograms of water.

All told, the water needs of the United States come to 2,700 cubic meters per year per person.

In those regions of the world where industry is negligible and where agricultural methods are simple, water needs can be satisfied by 900 cubic meters per year per person, and the average figure for the world as a whole might come to 1,500 cubic meters per year per person.

This looks hopeful. The average requirement of 1,500 cubic meters per year per person is a trifle under 1 per cent of the world supply, and even the lavish American use of water is well under 1.6 per cent of the world supply.

How can we possibly talk about a water shortage, then?

Easily! Let us reexamine our assumptions.

In the first place, liquid fresh water is *not* evenly spread among the population of the Earth.

In some places, it is superabundant and is present in greater quantities than human beings can use, or do use. As an extreme case, consider the Amazon River, which drains the wet, equatorial areas of South America. The Amazon is the largest river in the world and discharges into the sea in one year enough fresh water to supply 1,800 cubic meters for every person on Earth. That is enough, in itself, to supply all the needs of humanity if it could be saved and distributed. In actual fact, virtually none of the Amazonian water supply is used by man.

In other places, the liquid fresh water supply falls far short of average and leaves regions of the Earth arid or semiarid. In the past, such areas bore only that limited load of life that the water supply would support. The coming of human beings has, however, made a difference.

Advancing technology has enabled water supplies to be tapped from deep wells and distant rivers. Success in doing so has turned some naturally desert regions into garden spots with flourishing agriculture and industry and large populations. All this tends to increase to the limit allowed by periods of natural or high rainfall. But then, when the rains fail over a period of years and the drought comes (as must eventually happen, sooner or later), the garden spot finds it is still a desert after all, and there is a dire emergency—as in the western United States from time to time.

Another example is in the Sahel region to the south of the Sahara desert. There population has increased and water use became more lavish thanks to imported techniques of agriculture and of living. And then came a three-year drought in the middle 1970s which killed a hundred thousand people.

Then, too, human overexploitation of the environment has led to its deterioration. Unwise agricultural practices have created dust bowls and lost Earth's surface tremendous quantities of its topsoil. Herds of domesticated animals, goats in particular, have stripped land of its vegetation.

With topsoil and vegetation diminished, the land is less capable of absorbing and holding rainfall. The rain runs off more rapidly, accelerating the loss of topsoil, losing the land its fertility and creating and extending the deserts—a process which has now come to be called "desertification."

Finally, human industry, which uses the fresh water of rivers and lakes as a convenient sewer for chemical and thermal wastes, and an increasing population which fills them with biological wastes, all serve to pollute our freshwater supplies to a greater and greater extent and render them increasingly unfit to use.

Add all this together, and a water shortage right here on superwatery Earth becomes not only a future possibility but an unavoidable present fact.

Well, then, what can we do about it?

There are a number of possibilities.

1. Since the ultimate source of water is the ocean and since all the water evaporated from it returns to it eventually, it will do no harm if we try to speed that evaporation. By desalting the ocean one way or the other, we can obtain a direct and essentially unlimited supply of fresh water. To do so costs energy, however. Energy-rich, water-poor nations, such as Kuwait and Saudi Arabia, are obtaining fresh water in this fashion, but they are sparsely populated nations with limited needs. For ocean desalination to work on a large scale, we will need new energy sources that have not as yet been tapped.

2. Since rain falling on the ocean is completely wasted, anything that will encourage rain on land, at the expense of rain on the ocean (short of producing floods), is all to the good. Land can be built up at the expense of the ocean in certain favorable areas, as in the Netherlands, and can receive rain that would otherwise fall on water, but this can never be more than a very limited device. Cloud seeding, or still more sophisticated methods to be developed in the future, may succeed in directing rainfall precisely to those areas where it is needed. However, the benefit to one region will usually come about at the expense of another, so the ecological (and political) results might be unpleasant.

3. Since snow falling on the ice sheets is of no use to us, we might develop methods for exploiting the frozen fresh water of icebergs instead of letting them melt uselessly into the sea. We can imagine such bergs (especially from Antarctica) tugged to places like Chile, or even right through the tropics to the Middle East or to Los Angeles. It sounds a bit science fictionish, but it might be done. (The melting of the ice sheets themselves must be avoided under any conditions. If all the ice on Earth were melted, the water would drain into the ocean, raising the water level by sixty meters and drowning all the heavily populated coastal lowlands of the world.)

4. Since evaporation from lakes and from the soil is an important route for freshwater loss, methods for reducing evaporation might be developed. Single-molecule films of certain solid alcohols, or a layer of plastic balls, might be placed upon exposed fresh water to keep down evaporation. Israel, Chile, Italy, and the United States have experimented in this direction, but wind and wave tend to break up the inhibiting layer, and if the layer persisted there might be the danger of limiting oxygen renewal in the water below.

5. Since the river runoff into the sea represents a waste of fresh water, every effort should be made to make more efficient use of the water before allowing it to reach the sea. A complicating factor here is that all regions in the drainage basin have a claim on the river water; and different states, provinces, or nations might quarrel virulently and endlessly as to what the proper allocations might be.

6. Since the freshwater supply of the planet is very unevenly spread, water must be viewed as a regional and, ultimately, a global resource. Just as oil is produced in oil-rich regions and shipped to oil-poor regions, so water must be produced and shipped—perhaps in large plastic containers towed overseas by tugboats. (In the days before refrigeration, an important New England industry consisted of cutting ice out of frozen lakes and rivers and sending it elsewhere, sometimes shipping it considerable distances across the sea.)

7. Since groundwater supply is, in total, some forty times as great in volume as the water found in rivers and lakes, and is more widely spread besides, every attempt should be made to exploit that groundwater more efficiently—but not at a rate greater than it can be replaced by natural processes.

8. Since the supply of fresh water is increasingly critical, every effort should be made not to render any of what exists useless. This means that the pollution of lakes, rivers, and groundwater must, at all costs, be held to a practical minimum.

9. Finally, and above all, we must understand the limits of growth. As Earth's human population increases, the requirement for fresh water increases with it not only for drinking and washing, but for the agriculture needed to grow increasing quantities of food and the industry needed to produce increasing quantities of other commodities.

If population is not controlled, the continuing growth of the needs of humanity will eventually overtake any increase we can bring about in the freshwater supply (and in many other needs, material and spiritual); and the collapse that follows will be all the more catastrophic the longer it is staved off.

Given ingenuity, good sense, good will, and (most important of all, perhaps) good luck, we can yet create a flourishing and happy planet, but the time of grace during which we can accomplish this is now perilously short.

It is hard to imagine that salt is as essential to life as water is. Salting food seems to be an entirely voluntary act and to some people a "salt free" diet is beneficial.

Nevertheless, salt is essential to life and a discussion of it fairly belongs to a book dealing with life. When we avoid salt, all we do, and all we can do, is avoid excess salt.

13

SALT

The solid crust of the Earth is made up of a variety of substances, almost all of which are insoluble in water. This is fortunate, for it means that when the waters of Earth's voluminous ocean smash against continental shores, those shores are not dissolved. The continents do not melt into a slurry but remain intact so that life on dry land, including human life, remains possible.

There is an exception to this rule: one common material of the Earth's crust that *is* soluble, and that is salt—sodium chloride—NaCl.

("Na" is the chemical symbol for the active and dangerous metal, sodium, and "Cl" the chemical symbol for the active and dangerous gas, chlorine. When their atoms move into contact, an electron passes from a sodium atom to a chlorine atom and each is tamed in the process and turned into a bland and harmless substance. The two together form salt.)

Since sodium chloride is soluble, the rains wash the salt of the land into the sea bit by bit, and the ocean, as a result, is not pure water but is a salt solution or "brine." It contains other things than salt, but salt makes up the bulk of the dissolved matter of the ocean—more than three quarters of it.

On the whole, the ocean is 3.5 per cent dissolved solid matter. If all the salt in the ocean could be taken out and spread evenly over the land surface of the world, it would form a layer 475 feet deep everywhere. If the salt were confined to Manhattan Island only and were heaped up vertically, higher and higher, it would reach five-sixths of the way to the Moon. Its total volume is 4,500,000 cubic miles and its total weight is 47 million billion tons.

Clearly, there is more salt on Earth than mankind is ever likely to need.

We do need some, though, and the most intimate need is for the maintenance of life, all life. Salt is essential to the chemical processes that go on in living tissue.

Plants obtain the salt they need (along with other minerals) from the water in the soil, since that water dissolves any salt present. Animals require, for some reason, greater quantities of salt than plants do, and their tissues are correspondingly saltier. Carnivorous animals, eating other animals, are likely to get an ample supply of salt for their needs. Herbivorous animals, eating plants, are very likely to be salt-starved and must then find additional sources.

There are places where salt is to be found on the surface of the Earth, and herbivorous animals will flock to these places to lick the exposed salt. Such "salt licks" are as necessary, in the long run, to animals as food and water are.

Human beings are omnivorous and their salt needs depend on their diet. As long as a group of human beings obtain their food by hunting, or by herding, and eat quantities of roast meat and drink quantities of milk, they will get all the salt they need.

Once a changeover is made to agriculture, however, and grain becomes the chief item of the diet, things change. This is especially true if the advance of waterproof pottery makes it possible to boil meat, thus leaching the salt out of it. On such a diet, there would be salt starvation. It is the craving for salt that causes us to salt our food as we eat and find the resulting taste a great improvement. (Undoubtedly, the habit of salting food led to the discovery that the salting of meat tended to slow the process of decay and preserve it for periods of time. This increased the efficiency with which slaughtered animals could be consumed and was one more item among those that made it possible for the human population to rise.)

Such is the necessity for salt that innumerable expressions have come into use that refer to it favorably. It is used to represent whatever is good. Shakespeare speaks of the "salt of youth." Wit is referred to as the "salt" of conversation. Jesus addresses those he considers worthy as "the salt of the earth."

To doubt someone's word is to "take it with a grain of salt" since only salt will make it palatable. To partake of someone's hospitality, and thus owe him a debt, is "to eat of his salt." Since the salt container was in the middle of the table so that it could be available to all, those at the honored head of the table were "above the salt," the others, "below the salt."

To spill salt—lose something valuable—is clearly a stroke of ill fortune and a sign of bad luck. Soldiers were sometimes given salt, as they were given rations; this was called a "salarium" from the Latin word for "salt" and our word "salary" comes from that.

In earlier times, places where salt could be produced had the kind of economic power we now associate with oil-producing nations, and cities such as Rome and Venice laid the foundations of their greatness on their ability to control the salt trade of a region.

Obviously, one of the surest sources of revenue was a tax on salt, since people couldn't do without it. The French government had placed such a tax (the "gabelle") in the 1200s, and it was one of the impositions that particularly infuriated the populace. This fury helped bring on the French Revolution, and one of the first things to go under the new Revolutionary regime was the gabelle.

At the time of the French Revolution, the Industrial Revolution was also underway in Great Britain, and with industrialization, the need for salt multiplied many times. This was not so much because of the needs of an increasing population but because of its use in every facet of industry.

Salt—common, cheap, soluble, easy to handle—is the ultimate source of eleven basic chemicals which are, in turn, used as sources for other materials, so that salt can be found at the root of almost every existing chemical product. What's more, it is used for a wider variety of purposes in the various facets of human activity than any other mineral substance. It is estimated that there are more than 14,000 separate and distinct practical uses for salt, as solid or in solution.

Today, salt requirements are mainly for human consumption in the underdeveloped nations, but in a country like the United States, several times as much salt is used in industry as in food.

It is a mark of the increasing industrialization of the world that the production of salt has been rising rapidly in recent years.

Since 1960, the production has nearly doubled. In 1970, it reached a level of over 150 million tons per year, of which the American production is 45 million or nearly a third of the whole.

Even this vast production, however, represents but a tiny fraction of the salt content of the ocean—less than a hundred millionth—so in this direction, at least, we need fear no scarcity, particularly since the salt we use is eventually washed back to sea anyway.

The most obvious source of salt is the ocean, and the simplest way of getting salt out of the ocean (a procedure that requires energy) is to let the sun do it. If ocean water is placed in large, shallow pans, the sun will vaporize the water and leave its solid content behind.

This cannot be done profitably everywhere. High temperatures are required, a sun that is not too frequently hidden by clouds, air that is not humid, and rain that is infrequent. In hot, dry weather, evaporation is hastened, so that the production of "sea salt" can be carried through most easily on tropical or subtropical coastlines, fronting arid or semiarid areas.

Another difficulty is that one can't simply evaporate the seawater in one operation and consider the solid residue to be salt in any satisfactory way. Although three fourths of the residue is indeed sodium chloride, the remaining quarter is a mixture of magnesium chloride, magnesium sulfate, calcium sulfate, potassium chloride, magnesium bromide, and calcium carbonate.

None of these impurities is violently poisonous, but they are undesirable for one reason or another. They can add bitterness to the taste, and they possess a tendency greater than that of sodium chloride itself to absorb moisture out of the atmosphere. If the solid mixture obtained from seawater were ground, the grains would, except on very dry days, absorb water and stick together in a large clump.

Relatively pure sodium chloride has only a slight tendency to do this, and the tendency can be reduced even more if a very small quantity of magnesium carbonate or calcium silicate is added. Such salt, when powdered, will remain dry and uncoagulated even on muggy or rainy days and will move easily through the holes of a salt cellar. (Hence, the advertising slogan of one salt company: "When it rains, it pours.")

Early man did not, of course, know the composition of seawater, or the nature of the dissolved solids, or the rationale behind clever techniques for separating the sodium chloride from the other chemicals. By hit and miss, however, he learned the way to produce reasonably palatable salt.

The different substances dissolved in seawater are soluble to different extents. If seawater is slowly evaporated, those substances that are less soluble than sodium chloride settle out first. The seawater can then be transferred to another container while the initial crystals are discarded. Further evaporation brings down a residue that is mostly sodium chloride, and the rest of the liquid is discarded, since drying it further would merely add the more-soluble impurities. The salt produced is raked, drained, rinsed, dried, and finally powdered.

As seawater grows more and more concentrated, life in the water that is adjusted to a low-salt concentration dies out. Some forms of life that flourish in a higher salt concentration briefly multiply and predominate. Some salt-loving algae turn the water red, orange, and brown, as they live their short spans in an environment in which none can compete with them. Eventually, they die, too, but where salt is prepared too carelessly, some may live on and contribute to the spoiling of the meat that the salt is intended to preserve.

Even very primitive methods, however, will suffice to produce a low-grade but usable salt from the sea. In emergencies, it will do, and when nothing else is available, people are delighted to have it. During the Civil War, people in the Confederacy, starved of salt by the Union blockade, sometimes made salt for themselves out of seawater, and in the 1930s, Mohandas K. Gandhi led his Indian followers to the coast to manufacture salt in defiance of a British government monopoly.

On the other hand, beleaguered people far from the sea can be in serious trouble through salt shortage. The Chinese Communists, resisting Chiang Kai-shek from their inland strongholds, were more threatened by Chiang's salt embargo than by his armed forces.

In industrialized nations, the techniques of producing salt from seawater are mechanized. The seawater is heated artificially and its evaporation hastened by means of open fires or steam jackets. The number of stages of evaporation are increased, and so on.

Even today, nearly half the salt produced in the world is

obtained, with varying degrees of sophistication, by the drying of ocean water in the sun.

The ocean is not quite uniformly salty. Where the water is cold and evaporation is slow, and where fresh water, either from flowing rivers or melting ice, is being added in quantity, the solids concentration is considerably less than normal. Where tropical parts of the ocean are trapped between desert shores so that evaporation is great and the addition of fresh water minimal, the solids concentration is higher than normal. The solids concentration in the Red Sea is nearly 5 per cent, and salt production there would be that much the easier.

There are also places on Earth where water collects into large pools that have no connection with the ocean, and where evaporation will have a greater effect on the smaller total quantity of water. The result is that inland seas can have a far greater salt percentage than the ocean itself. They would, in effect, be natural evaporating pans much larger than any that could be constructed by human beings, and, in them, the water has already partially evaporated.

The best example of such an inland sea is the Dead Sea (so-called because its high salt concentration does not allow a content of living things) which lies on the border of Israel and Jordan. The water of the Dead Sea is some seven times as rich in dissolved matter as the ocean is.

The only other comparable salty body of water of sizable extent is the Great Salt Lake in Utah, which has four times the surface area of the Dead Sea (1,500 square miles for the former, 370 square miles for the latter). The Great Salt Lake is shallow, however, while the Dead Sea is unusually deep, so that the latter contains the greater quantity of water. There are 75 cubic miles of water in the Dead Sea and over 12 billion tons of solids, with the Jordan River bringing in 850,000 more tons of solids each year.

Only a little over a quarter of the dissolved material in the Dead Sea is salt, however. The nature of the soil through which flow the waters feeding the Dead Sea is such that nearly half the solids content of the water is magnesium chloride.

Salt can be more easily obtained from such a semievaporated puddle. The solids even settle out spontaneously from the spray which occasionally coats objects near the shore of the Dead Sea, producing white clumps of impure salt that must have given rise to the legend of Lot's wife turning into a pillar of salt as she fled the Dead Sea city of Sodom.

While salty puddles as large as the Dead Sea, or even the Great Salt Lake are rare, there exist briny places—salt marshes on the surface or brines that exist at considerable depths. Some wells bring up brine that contain solids of the same mixture present in seawater but in much greater concentration. Other brines contain solids that are almost entirely sodium chloride, and salt from such a source is generally far superior in quality to sea salt—but is less common, of course.

There are brines that, like the water of the Dead Sea, contain considerable materials other than salt, and sometimes substances not usually found in dissolved form, such as strontium chloride and barium chloride. Such substances obtained as a by-product of salt manufacture can have special importance to the chemical industries—and may even be sought for their own sake, quite apart from salt.

There are also rich salt sources found on land.

This may seem strange at first thought. After all, the ocean is billions of years old, and it has served as a continual reservoir of water vapor which, as freshwater rain, falls on the land and trickles through the soil, dissolving salt and carrying it steadily back to the ocean. It should seem, perhaps, that by now all soluble material in the soil has been brought to the sea once and for all.

Were that true, land life would be thin indeed, for it would be starved of essential minerals. Fortunately, there are natural processes that restore salt to the land. For one thing, storms whip water droplets from the ocean far inland and, as they evaporate, a dust of erstwhile water-dissolved solids settles over the land's surface.

It is from the sea in this manner that traces of iodine are deposited on the land. These find their way into vegetation and supply animals (including ourselves) with the iodine that is essential for the working of the thyroid gland. To avoid trouble in areas where the soil is iodine-short, small quantities of potassium iodide are added to salt and this "iodized salt" is now a common article of human diet.

A much larger sea-born resource is the salt that arises from the shallow ocean inlets that existed here and there, on what is now the continental surface, in ages past. Such inlets must occasionally have been pinched off into super-Dead Sea objects and must eventually have dried up, leaving thick deposits of salt behind—an example of natural evaporation on an enormous scale.

Such deposits are, of course, subject to being washed back into the sea again by the rain except where they are located in arid regions or where they are, through various geological processes, buried deep enough to be unaffected by rainfall. Sometimes these layers of salt are many hundreds of feet thick and, large or small, are found in many parts of the world. In the United States, for instance, twenty-eight states have substantial salt resources within their boundaries.

Where salt lies fairly close to the surface, either because it has formed there, or because pressures within the earth have forced deep layers upward into "salt domes," they can be used as a source of salt. It doesn't require much to be able to hack up slabs of the white material.

Where salt layers are deep underground, they can be mined much after the fashion of coal. Shafts are pushed down from the surface to the salt, and corridors are then driven horizontally through the strata, the salt being drilled, blasted, broken up, ground, brought to the surface as "rock salt," and often not further refined.

The salt layers have varying compositions, depending on the manner in which the seawater happened to have evaporated. There are places, as for instance in Poland and in the states of Michigan and Louisiana, where, through some lucky combination of circumstances, almost pure sodium chloride has been laid down.

However, even if the rock salt, as brought up, is not pure enough to be entirely palatable for human consumption, it may still do for many industrial purposes. The salt used to break up and melt ice on winter streets and highways in the northern states need not be very pure, for instance.

Rock salt need not be mined directly by sending men down after it. The fact that it is soluble means that water can do the fetching. Once a shaft is driven down to the salt layer, water can be pumped down through it and eventually be brought up as brine. Thus in itself is a purifying procedure since admixed soil, or substances only slightly soluble, will be left behind.

Once the brine is brought to the surface, it can be evaporated in the same fashion that seawater is except that the product is usually more nearly pure sodium chloride.

The brine can be evaporated in long shallow pans, as much as 150 feet long and capable of holding as much as eighty tons of brine.

The brine can be heated artificially to hasten evaporation, and evaporation can be hastened still further if the heating is done under a partial vacuum, for as the air pressure is lowered, the boiling point of the brine drops. The brine is shifted through stages of increasing vacuum in a cycle taking up to forty-eight hours, with new brine introduced at one end as the salt crystals are removed at the other.

The brine is sometimes initially heated in closed containers to high pressures and then passed through a large vessel filled with stones on which some of the less-soluble impurities settle out readily, allowing the brine to trickle through with a purer salt content. The pressure can then be gradually reduced, and when the salt begins to crystallize, the brine can be placed in large shallow pans, where further evaporation will precipitate more salt.

But however salt is produced, whether from the sea or on land, whether by the action of solar or artificial heat, whether with or without sophisticated engineering techniques, it remains an all-round essential substance for life and industry—and one essential resource of which we can never run short.

I can almost follow the events of the day by the type of requests I get for articles. If a landing on Mars is in prospect, I am quite certain that I will get a rash of requests for articles on Mars.

Since the middle 1970s saw a number of sizable earthquakes, I suspected there would be a feeling that Earth was shaking itself to pieces for some spectacular and mystical reason. I knew these theories of destruction would be espoused with particular ardor by people who knew nothing about the subject.

This meant, I was sure, that I would get requests for articles describing earthquakes from the standpoint of science—and I wasn't disappointed.

When I was asked to take up the subject by an unexceptional periodical such as Think, *the house organ of IBM, I seized the chance and the following essay was the result.*

14

EARTH SHRUGS ITS SHOULDERS

What natural disaster comes without warning and can kill a million people in five minutes? —An earthquake.

If your answer was a tsunami (more commonly, and mistakenly, called a "tidal wave") that is about the same thing. A tsunami is initiated by an earthquake centered beneath the ocean floor.

All who have experienced a major earthquake (and have survived) agree that the terror it inspires is unexampled.

It seems to violate the course of nature. You expect the rain to fall, the rivers to rise, the summer to be hot, the winter to be cold. Excesses or deficiencies of these may be too much or too little for comfort, but this is expected, sometimes, too. And you can escape. You can find shelter. You can reach high ground.

But an earthquake? It is the solid ground itself that shakes and heaves as Mother Earth shrugs her shoulders. Who expects the eternal solidity of the Earth itself to give way? And when it does, there is no place you can go. All shelters become deathtraps. You can only wait until the ground becomes solid once more—and then can you ever trust it again?

Those who have not experienced an earthquake rarely give it any thought at all. If there are earthquakes somewhere, why they are elsewhere in space, or in time, or both.

Until this year—1976.

Suddenly this year, we were bombarded with disasters in Guatemala, in Mexico, in Italy, in China. Everywhere, the ground was trembling. Everywhere, the houses came down. Everywhere, the people were camping in the streets or fleeing.

What's happening? Is the earth shaking more and more? Has something happened to our planet? Is it falling apart?

Probably not. As in the case of every other natural phenomenon, incidences vary from year to year. There are some years in which earthquakes are more numerous and damaging and some in which they are less so. This past year is probably well within the usual boundaries for this geologic era.

Why, then, do things *seem* so bad. For this, there are three reasons.

First, communications have improved enormously since World War II. It was not many years ago that vast areas of Asia, Africa, and even South America were out of touch with us. If an earthquake took place in a remote region of Turkey or Iran or China or Chile, only the dimmest word reached the American public. There might be a small item on an inside page of the newspaper headlined "Quake in China," but then nothing else would ever be heard of it. Now, every earthquake is described at once and in detail on the front pages. The results can even be seen on television.

Second, our own interest has grown. We are no longer isolated and self-absorbed as we were not so many decades ago. There was a Denver newspaperman who once said that a dogfight in a Denver city street was more interesting to his readers than an earthquake in China was, and he was right. Even if quakes were reported then, nobody cared—unless it happened to be in California, and then California cared. Nowadays, however, the world has grown small and Americans have learned that almost any event anywhere is likely to affect them at home—so they have grown interested, and pay attention.

Third, the world population has grown. It has doubled in the last fifty years and stands at four billion. The cities have grown even faster than the countryside so that not only are there more people, but on the whole, they exist in larger and denser concentrations. They often inhabit rickety houses, tens of thousands of them, row on row, which are not designed to withstand a ground shudder. Add to that that the works of man are more complex by far and more costly than they ever have been in the past. Therefore, when an earthquake does take place now, it is quite likely to kill more people under collapsing houses and do far more property damage than the very same quake would have done fifty years ago—so earthquakes *seem* to be growing worse.

The San Francisco earthquake of 1906, for instance, which was the worst the United States has yet experienced, lasted forty-seven seconds, killed four hundred people, burned out four square miles with the fire that followed, and did half a billion dollars worth of damage altogether. Imagine a quake of the same intensity and duration hitting San Francisco today, and imagine the city as unwarned and as unprepared as it was in 1906. The enormous damage—I don't even dare estimate what it might be.

What causes earthquakes?

The ancient Greek myths had it that they were caused by rebellious giants whom Zeus had imprisoned underground. Occasionally, the giants, chafed by their chains, shifted position and the ground trembled.

Greek philosophers, scorning the supernatural, suggested, on the other hand, that air was trapped underground and that it was the trapped winds that made the earth shudder now and then.

Modern science could not do much better than that for a long time, though it did learn to detect and measure earthquakes with delicate exactness.

In 1855, the Italian physicist, Luigi Palmieri, devised the first workable seismograph, an instrument based on bedrock with a pen (nowadays a ray of light) capable of producing a wiggling line as it shook along with the shaking earth. When several seismographs, widely spaced, detect the same earthquake, its center and intensity can be determined.

In 1935, the American seismologist, Charles Francis Richter, devised the "Richter scale" as a measure of earthquake intensity. The intensity is given as the logarithm of the maximum displacement indicated by the seismograph at a fixed distance from the center of the quake. Each number represents an intensity that is a fixed number of times greater than that of the next lower figure.

On this scale, the number 2 represents a barely perceptible shock, while 6 represents an earthquake capable of doing important damage. Anything over 7 is a major earthquake.

The Guatemala quake of February 4, 1976, was 7.6 on the Richter scale. The San Francisco earthquake of 1906 and the near Peking earthquake of 1976 were both 8.2 on the Richter scale. Nothing higher than 8.9 has ever been measured, and such a top-rank earthquake would release the energy equivalent (minus radioactivity and heat, of course) of one hundred large H-bombs.

But what *causes* earthquakes? The answer came only in the 1960s through studies of the ocean floor, to begin with, and especially of the great mid-ocean rift.

The Earth's crust is not a single piece but consists of large plates in contact. In some places along the joints between plates—indicated by the ocean rifts—hot material from the Earth's depths slowly wells upward, forcing those two plates apart.

In other places, as a result, two plates are pushed against each other. Where the plates meet, there can be buckling and mountain formation; otherwise one plate will slip downward under the other, moving into the high temperatures below and melting.

Where two plates are forced together, there is the possibility of sidewise slippage. The vast pressures of the two plates against each other produce great frictional resistance to such slippage, so the plates remain motionless while the forces producing the slippage increase and increase until, quite suddenly, they overcome the frictional resistance. There is then a sudden movement and the vibrations of that movement are the earthquake.

Most earthquakes take place in the vicinity of these joints between plates—in a rim around the Pacific, for instance, and in an east-west line across the Mediterranean and central Asia. There are minor "faults" outside the plate joints which occasionally give rise to quakes, too, both less common and less powerful than the others.

In 1976, the African and Indian plates apparently moved northward perceptibly (a matter of a couple of centimeters) and that seems to have touched off shocks all along the northern border of the plates from Italy to the New Hebrides islands.

Can we predict earthquakes?

In theory, we can. As the pressure, which will cause one side of a fault to slip against the other, increases, some minor changes must take place in the ground prior to the actual shock, and these must, in some way or another, be capable of being measured.

There are changes, for instance, that are referred to as "dilatancy." As pressure on the subterranean rocks builds up, tiny cracks in the rock are pulled apart and dilate. There isn't enough underground water to fill those larger cracks just at first, so the overall density of the rocks decreases slightly. That means that vibrational waves passing through the rock do so at a less than normal speed. But then water seeps into those larger cracks and fills them so that the vibrational waves speed up again and by that time the fault is about to give.

Then, too, the changes in rock as it begins to give just before an earthquake include a decrease in the electrical resistance, a humping upward of the ground, and an increase in the flow of water from below. The increased flow of water can be indicated by an increase in the trace of radioactive gases present in the air, gases which have till then been imprisoned in the rocks. There are also rises in the level of well water and an increase in muddiness.

Oddly enough, one of the important signs of an imminent earthquake seems to be a general change in the behavior of animals. Normally placid horses rear and race, dogs howl, fish leap. Animals, like snakes and rats, which ordinarily remain hidden in holes, suddenly surge into the open. In California, behavioral scientists, monitoring chimpanzee behavior, find that they turn restless and spend more time on the ground than usual a day or two before a quake.

We don't have to assume that animals have the ability to foretell the future or have strange senses we don't have. They live in more intimate contact with the natural environment and live precarious lives that force them to pay more attention to tiny changes than we ever do. Tiny tremblings that precede the real shock would upset them; strange sounds arising from the scraping of the lips of the fault would do the same.

In China, where quakes are more common and damaging than they are here, great efforts are being made at predicting quakes, and the population is mobilized to be sensitive to change. Strange actions of animals are reported, as are shifts in the level of well water, the occurrence of strange sounds from the ground, even unexplained flaking of paint. In this way, the Chinese claim to have anticipated damaging earthquakes by a day or two and saved many lives—notably in the case of a quake in northeastern China on February 4, 1975, which was 7.3 on the Richter scale. (On the other hand, there is no indication that the Chinese were in any way prepared for the monstrous quake south of Peking in 1976.)

In the United States, too, attempts at earthquake prediction are becoming more serious. Our forte is high technology and this is the direction in which we will probably turn. The delicate detection of changes in local magnetic, electrical, and gravitational fields could be useful, for instance, as would be the day-to-day changes in the level and chemical content of well water, the properties of the air, and so on.

It will be necessary, however, to pinpoint the possibility of an earthquake occurrence quite accurately. The effort at rapid evacuation might do more in the way of economic dislocation and personal discomfort than a minor earthquake would, and if the earthquake *were* minor or didn't come at all, there would be a net loss. Besides, if one warning turned out to be a false alarm, the next warning might be disregarded, and that time a major shock might come.

Probably, to increase the chances of predicting an earthquake with reasonable certainty, a variety of measurements would have to be made and the relative importance of their changing values weighed. One can well imagine that the quivering readings of a dozen different needles, each measuring a different property, would be fed into a computer, which would constantly weigh all the effects and yield an overall figure which would, on passing a certain critical point, signal evacuation with a high probability that it would not be a false alarm.

Can earthquakes be prevented? There is no practical way in which we can modify subterranean rock, but subterranean water is another matter. If deep wells are drilled several kilometers apart along the line of a fault, and if water is forced into them and then allowed to backflow, subterranean pressures might be relieved, and an earthquake may be aborted.

In reverse, it might be possible to encourage minor shocks at intervals. A group of minor shocks, spaced in time, would release as much energy, in total, as a major shock would, but the minor shocks would do so little by little and do no damage in the process.

There's hope.

One problem with science writing is that, with inconvenient rapidity, essays may go out of date in one way or another. One way of correcting this is to seize the chance to write a second article in which the material can be updated. There is then bound to be some overlapping, of course, but it is worth doing.

In 1966, for instance, I wrote an article on memory for The New York Times Magazine, *one which appeared in my collection* Is Anyone There? *under the title of "I Remember, I Remember."*

Five years later, Penthouse *asked me for an essay and I seized the chance to write another one on memory in order to incorporate new findings I thought were important. Here it is.*

15

FORGET ME NOT!

We talk a great deal about studying the infinitely vast and the infinitely small. Astronomers deal with quasars that are eight billion light-years away, and physicists deal with resonance particles that live for a trillionth of a trillionth of a second. We stand in amazement at such discoveries, yet these sink into insignificance compared with the real wonder of the universe that is part of ourselves.

Each of us carries three pounds of material that is far more complicated than anything else that scientists study. The distant quasar and the tiny resonance particle are more apt to be analyzed in satisfactory fashion, soon, than is the soft, grayish mass of material within your skull.

For complexity, there is nothing like the human brain. The brains of elephants and whales are larger, but on the basis of the evidence of things accomplished, the human brain stands without a peer.

The human brain contains about ten billion nerve cells and a hundred billion smaller auxiliary cells. Each cell is extraordinarily complicated and is equipped with tiny branching filaments extending outward in all directions. These living filaments approach each other so that those from two neighboring cells are separated only by tiny gaps called "synapses." Across those synapses, communication can take place by way of chemical molecules and electrical impulses. Each cell can make such connections with up to dozens of its neighbors.

It is the complexity that results from billions of cells in a set of intricate connections that makes it possible to learn, reason, imagine, and create on the human level. And every aspect of mental activity rests on memory. To learn means to gain new memories. And it is on the basis of these memories, old and new, that we do everything else—even create, for no creation is utterly new but invariably stands on the basis of the old and remembered.

But what is memory? Somehow every sense impression to which we are exposed leaves its mark upon the brain—a mark that remains for a longer or shorter time. And somehow, by an effort of will, we can bring up from all these imprints something that is relevant to our immediate purpose.

Few of us are satisfied with the efficiency of our own memory mechanism. Few of us feel that we remember all we should or as quickly as we should; and so we forget the really remarkable feats we accomplish even with a "poor" memory. The mere fact that we can speak reasonably well means that we can remember thousands of words, recall each one as we need it, and recall, too, something of the system of putting them together in such a way that others can understand what we are saying. This, alone, is something nothing else on earth, but the human brain, can do.

How does a particular word insert itself into the memory; the word "brain" for instance? If I were to ask you the name of the object inside a person's skull, you would say "brain" at once, but how did you select that sound from all the different sounds you know the meaning of? Did you somehow check over all the sounds you know and pick out the right one? Did you associate "brain" with "skull"? As soon as you heard the latter, did you think the former? But I might have asked what the skull was composed of and you would have said "bone."

Or suppose I had asked you if "blain" was an English word. You would probably say no at once. "Brain" and "plain" are

English words, but not "blain." How could you tell? Did you check over all the English words you know and notice that not one of them was "blain"?

These are very simple cases of memory, but they are already enough to stump scientists in the field. The sad fact is that no one knows how human beings remember. No one knows how human beings recall what they remember.

What does seem obvious, though, is that the capacity to remember is enormous. Suppose we consider the smallest unit of fact, a yes-or-no item, to be a "bit." What seems a single memory may involve a considerable number of bits. For instance, if you can remember your father's face, you can remember whether his eyes are blue or not blue; whether he is bald or not bald; whether his face is broad or not broad; whether his lips are thick or not thick. Each memory, and many more about that one remembered face, is a bit. Some estimates have placed the total number of bits a brain can acquire in the course of a long lifetime to be a quadrillion: 1,000,000,000,000,000.

On the average, then, you accumulate ten trillion bits a year and from these bits you have to recall the particular bits you want. If you are ever impatient with yourself for forgetting, then, remember to spend a little time marveling at the fact that you remember anything at all.

But how can so many bits possibly be stored? To imagine that each brain cell is, for instance, devoted to one bit is ridiculous on the face of it. There may be more brain cells in your head than there are people on the planet, but that number still isn't enough. The memory system must be something much more subtle.

Suppose that a specific memory is not stored in a single cell but that it consists of a pathway from cell to cell. Imagine ten brain cells, each one of which is connected by delicate little fibers to all the others. A tiny electrical impulse can leap the synapses and pass from the first to the second to the third, and so on, until the tenth is reached.

But it can do this by a variety of paths. It can go 1-2-3-4 . . . or 1-3-2-4 . . . or 3-4-1-2 . . . and so on.

The total number of different ways in which an electric current can pass through each one of ten cells after starting from any one of the ten is, actually, 3,628,800. (If eighteen cells were involved and if each was connected to each of the other seventeen, then the total number of different pathways an electric current could take to pass through all of them would be about 6,400,000,000,000,000. If each pathway could represent a single bit of information, then the total number of pathways would be more than six times as great as the total number of bits a human brain could accumulate in a lifetime.

One could imagine a small complex of eighteen cells holding a lifetime of memories. Naturally, we would not expect the brain to work in this fashion. Too much would be concentrated in too little. A tiny injury might wipe out all memory. It is much more likely that the brain works with a great deal of leeway for error. Memories might be duplicated a thousand times over in different parts of the brain. There would be room for it. A thousand different groups of eighteen cells, placed strategically here and there in the brain, each one with an independent store of memories, is conceivable at any rate.

Or suppose that instead of different circuits, there are molecules involved. In each of the billions of cells in the brain there are many billions of molecules. Some of these are the rather large and complex molecules of proteins, which are the most versatile materials of living tissue. There are at least a million times as many protein molecules in the brain as there are total numbers of bits in man's longest, largest memory store.

Could there be a different protein molecule for every bit?

Each protein molecule is made up of a string of amino acids of about twenty different varieties. An average protein molecule might be made up of three hundred amino acids, each one being any one of the twenty. The number of possible different combinations of amino acids is far beyond the merely astronomical. If every protein molecule in every living creature that ever lived during Earth's whole existence were different, it would be a mere pinprick to all the different protein molecules that could exist.

A chain of merely fourteen amino acids could exist in enough different varieties to account for four times as many bits as the human brain can gather in a whole life. If every one of those bits was represented by a different fourteen-amino-acid molecule and the whole thing were repeated twenty times over in different parts of the brain, it would only take up a millionth (at most) of all the protein in the brain.

Memory seems to be divided into two kinds: short term and long term. If you look up a phone number, you will repeat it over to yourself and you will then remember it long enough to dial it. By the time you complete your telephone conversation, you may have forgotten it, and you may never remember it again.

If you have occasion to use the phone number periodically, you may find, eventually, that you don't have to look it up. It is now part of your long-term memory. In many cases, items in the long-term memory can be recalled without very much trouble even after years have passed in which you have not had occasion to use the memory.

Suppose short-term memory involves the cell combinations? Perhaps the act of looking at a phone number somehow lowers the resistance at certain synapses so that a small current can travel easily along a certain cellular route. As you keep the number firmly in mind, the small impulse continues to spark from cell to cell along that definite route.

If you keep it up long enough, or have occasion to go back to the number periodically, molecules may perhaps be formed and then the memory becomes long term. On the other hand, if you use the memory only briefly and then turn your attention to other things, the synapses slowly restore themselves to their original state and the memory is lost.

Perhaps the molecules of long-term memory only have a certain lifetime, too. Perhaps every time you use a long-term memory, more molecules representing it are formed, and the longer the memory lasts. Perhaps if you leave even the strongest long-term memory alone for a long enough interval, all the molecules fade out, changing structure randomly and no longer serving as memory. In time you may forget the name of your first sweetheart.

Or do you? Do you really lose memories altogether or does every impingement on your brain leave a permanent trace, even those casual impingements you would think are just good for short-term memories? Is it that you block things instead of forgetting them? There are, after all, constant reports of people who remember things under hypnosis that they had utterly forgotten in a normal state.

Wilder C. Penfield, at McGill University in Montreal, has actually stimulated such memories physically. While operating on a patient's brain, he accidentally touched a particular spot that caused the patient to hear music. He could repeat this over and over. The patient could be made to relive an experience in full, apparently, while remaining quite conscious of the present. It was like playing back a tape recording. With other patients, he could make trivial snatches of memory come back, a person's voice, a casual action; often, it was music.

It seemed a purely random tapping of a memory store.

But if all is present and much is blocked, perhaps it is that the more a particular memory is used, the weaker is its block and the more amenable it is to further recall. Perhaps recall is rapid because you go through memories in the order of blocking. The thinnest blocks are first handled and usually you come upon what you want quickly. It is only when you need something you use only rarely that you must dive down into the thicker blocks, and that takes longer.

Of course, nobody knows of what the block is composed, how it works, and how one breaks through when one suddenly remembers something that seemed to have been forgotten.

How are we to penetrate all these possible complexities in which the only thing we can boast of, apparently, is total ignorance? We can take the attitude that in any learning process, in any storing of new memories, that is, *some* change takes place in the brain, and we ought to be able to detect it—whatever it is.

A Swedish neurologist, Holger Hyden, at the University of Gothenburg, developed techniques that could separate single cells from the brain and then analyze them for a chemical called ribonuclei acid (usually abbreviated as RNA). He subjected rats to conditions where they were forced to learn new skills—that of balancing on a wire for long periods of time, for instance—and, by 1959, found that the brain cells of rats so forced to learn showed a 12 per cent increase in RNA over the normal.

(The process of learning the intricate control of muscles is particularly rewarding. As children, we painfully learn how to use our muscles in such a way that we can ride a bicycle or skim along on ice skates and then, years afterward, with no practice in between, find we still remember.)

As it happens, RNA is the molecule that controls the formation of protein molecules in cells. Perhaps the learning process requires the formation of numerous special protein molecules, guided by the numerous special RNA molecules formed in the process of learning.

In the 1960s, Hyden discovered a particular type of protein in the rat brain. He called it S100, and it seems to occur *only* in brain cells. Its quantity increases during the process of learning.

The theory that the process of learning somehow brings about the formation of special RNA molecules which in turn guide the formation of special protein molecules gets support in other lines of experimentation.

Suppose, for instance, one uses a drug that interferes with the chemical process whereby RNA forms protein. At the University of Pennsylvania School of Medicine, Louis B. Flexner and his wife, Josepha, conditioned mice in a simple maze, teaching them to follow a particular path to avoid a shock. The conditioned mice were then given an injection of a drug preventing protein synthesis and promptly forgot what they had learned. Apparently, they were still only in the short-term-memory stage and now could not transfer it to the long-term.

If the Flexners waited five days before injecting the drug, it no longer had effect. By that time, apparently, enough protein had been formed to make the memory permanent. If the animal is trained first in one respect and later in another and the drug is administered, the later training is forgotten, not the former.

On the other hand, certain drugs tend to promote RNA formation and there are reports that these, on some occasions, tend to increase the speed of learning.

Does S100, or any protein formed in the process of learning, act as a memory molecule? Does it represent a memory bit simply by existing? Or does it *do* something? Is it the protein structure that counts, or the protein function?

Hyden, for instance, suspects that the S100 has some effect on the brain-cell surface. Perhaps it adds on to the membrane of the cell and makes it a little more apt to conduct the nerve impulse. Perhaps every protein molecule formed helps set up a particular variety of electrical circuit.

There is an important difference between a circuit representing a memory bit and a molecule representing it. A circuit is not a material thing so much as a relationship among cells. It cannot be transferred unless a group of cells is taken out alive and replanted alive in another brain.

A molecule, on the other hand, is a material thing and a nonliving one at that. It might well be taken out of one organism and injected into another.

And yet, just to make things more difficult, isn't it possible that each organism invents a protein or a circuit for itself every time it needs to store a bit of memory? In that case, transfer is impossible because each brain would have its own language that would be gibberish to any other.

Still scientists were bound to experiment. In 1961, James V. McConnell, at the University of Michigan, reported startling experiments with small organisms, called planaria, which are very low in the scale of life complexity. He subjected them to a flash of light and then to an electric shock. Their bodies contracted at the shock and eventually began to contract as soon as the light shone. They had learned that the light meant a coming shock, and that might mean the production of special memory molecules.

McConnell then chopped up the trained planaria and fed them to untrained planaria. He found that the untrained planaria, after their cannibalistic diet, learned to react to the light faster than ordinary planaria did. Had they incorporated some of the special memory molecules and obtained a headstart?

It was difficult to work with planaria, however, and to interpret their behavior. Not everyone accepted McConnell's results.

In 1965, the Danish physiologist, Ejnar Fjerdingstad, moved much farther up the scale and began to work with rats. He trained rats to go to the light in order to get food. He then collected the brains of a number of such trained rats, mashed them up, and injected the material into untrained rats. He found that such injected rats learned to go to the light very quickly. Somehow a memory of light-associated food had been transferred with the brain material.

The Hungarian-American physiologist, Georges Ungar, went even farther. In 1970, he subjected rats to an electric shock in the dark so that they finally developed a strong fear of the dark. Brain extracts, when injected into unshocked animals, caused them to show fear of the dark, too. From several pounds of brains from animals trained to show fear, Ungar isolated a chemical compound which would induce the fear in an untrained rat. What's more, it induced the fear in mice, too, and even in goldfish.

Not only were the memory molecules in one organism not gibberish to another organism of the same species, they were not gibberish even to organisms of another species. Ungar called his compound "scotophobin" from Greek words meaning "fear of dark."

It turns out that scotophobin is a very small protein molecule, made up of a string of just fourteen amino acids. What it does in the cell and how it works are as yet completely unknown, but it is the closest approach, so far, to anything that might be considered an actual memory molecule.

Ungar intends to go farther. He plans to train rats to become habituated to a loud noise and learn to pay no attention to that. He will then see if he can find a molecule that will transfer the opposite of fear. Working with tens of thousands of goldfish, he hopes to train them to distinguish blue from green (to go to a blue light for food, for instance, but not to a green light) and see if he can isolate a chemical for color distinction.

If several different molecules can be isolated, each with its own memory symbolization, some interesting generalizations might be drawn.

And what will all this mean for the future? As scientists learn more about the mechanism of learning and memory, will it mean that they will be able to improve the action of the human brain? Will they find conditions that favor rapid learning and efficient memory? Will they learn how to tap memories at will? Will they learn to unblock them permanently so that it will be possible for human beings to remember *everything?* Will all this raise the intelligence of human beings sky-high and make supermen out of us?

Perhaps we shouldn't hold our breaths over the possibility.

For one thing, the human brain is exceedingly subtle, and it may be a long time yet before these first gropings into the workings of that organ will yield firm results. And even after it does, there is no certainty that it would be wise to tamper with the brain.

Almost everyone is sure, for instance, that he would be better off with a better memory and that a better memory would somehow mean a better intelligence.

But would he and does it?

Some great scientists and mathematicians had prodigious memories, but others had very poor memories. What's more, some people with virtually photographic memories, who seem to forget nothing, are of only ordinary intelligence or less.

Are we even sure we want to remember better than we do? Perhaps there would be disadvantages involved?

Earlier in the article, for instance, I asked if "blain" were an English word and said it wasn't. Undoubtedly most readers agreed with me at once. A few, however, might have dimly recalled coming across the word "blain" in the Bible (see Exodus 9:9–10 and might have hesitated. Whether they finally decided yes or no, the point is they would have hesitated.

Can it be that too good a memory might result in hesitation

now and then in the course of life, where those with more efficient forgetteries would not hesitate? Would remembering too much get in the way of action? Would it get in the way of decision? Would it get in the way of further learning?

It might even be that the force of natural selection over millions of years of human evolution has worked against a too-efficient memory. Perhaps we fuzzy-minded individuals are the end product of a careful winnowing process. Perhaps we're the ones that survived best in the long run.

Let us not, then, be too anxious to improve a brain in a way that might not be an improvement. Mankind is learning these last few years that not all technological advances are necessarily beneficial or useful in the long run. Wherever we go and in whatever direction we move, we want to be careful that what we do does not backfire or lead to unexpected side results.

And where must this caution be most clearly expressed but in the case of a mechanism so delicate and complex that it has no fellow in the known universe?

If we are going to fiddle with the human brain, it can only be hoped that we do so with the greatest possible care. If there is anything we must *not* forget, it is that.

Notice that at the end of the preceding essay, I expressed some doubt as to the wisdom of too much tampering with the controlling mechanisms of life. I did the same in the following essay, which was published even earlier and which dealt with an aspect of life that is even more fundamental.

Incidentally, I must warn my Gentle Readers every once in a while that my cheerful self-satisfaction is no evidence, in itself, that everything I write is routinely accepted, even at my time of life and with my reputation.

The following essay was, for instance, written at the request of Playboy, *but once they had a chance to read it, they decided they wanted changes of a kind I considered unacceptable. I therefore retired the article and waited for an appropriate request from someone else.*

Such a request is bound to come along, although it may not be a lucrative one. In the present case, Boston University Journal *didn't pay me what* Playboy *would have. In fact,* Boston University Journal *didn't pay me at all, but on the other hand, they were willing to publish the article as I thought it ought to appear and that's worth more than money—at least if you're fortunate enough not to be short of money in the first place.*

16

YOU ARE A CATALOG

"I can read you like a book!" Thus says someone who finds another's motives transparently clear. It is a metaphor only, of course, but in the fast-moving scientific advances of these days, metaphors have a disconcerting habit of becoming sober fact. We *are* books, in a manner of speaking, and biologists are learning to read those books. To be more specific, every living creature contains a catalog, often in many copies. An individual human being contains millions of copies.

Within each catalog is a description of all the key parts an organism is likely to need in the course of "living" plus a system for ordering particular parts when needed and shutting off the supply when no longer needed. Different portions of the creature are supplied with different groups of parts, so that food is digested here, light detected there, poisons eliminated yon. The catalogs differ in size. They are slim indeed in viruses; so slim that the virus cannot quite make do on its own supplies. It must invade larger creatures and raid their better-supplied stocks. The catalogs are larger, but still small, in bacteria; larger still in more complicated organisms; and vast and encyclopedic in higher mammals such as man.

But all the catalogs, from virus to man, are (biologists have strong reason to believe) written in the same language. By removing the pages of the catalog from simple creatures and learning to read them, we learn at the same time to read the human catalog itself.

Once we learn to read the catalog, we may learn to order parts to suit our own purpose, rather than be forced to wait for organisms to do it in a way that suits theirs. We culture molds to make penicillin, for instance. Why not extract a particular catalog part and have it done without the mold? A synthetic system, minus the mold, may require less attention, and work with greater single-mindedness and, therefore, efficiency.

Perhaps by combining parts in new ways, or by deliberately altering some parts, we can manufacture substances not found in nature, and do so more cheaply and in greater quantities than our synthetic chemists can manage. By manipulating the catalog parts in an intact organism, we might perhaps make it perform feats beyond its ordinary power. A starfish can regrow a lost arm, while a human can't. Yet the human structure possesses the ability to do so at one stage of its development, for it grows two arms as an embryo. Can a portion of the remaining stump of a lost arm be removed, "reeducated" in the proper use of the catalog, and then restored in order that it might grow the new arm?

Similarly, can an affected gland be reeducated to produce a hormone in the proper manner? Can the body be reeducated to wipe out a number of diseases which, like diabetes as the most common example, arise out of an imperfection in the catalog?

There are some biologists who suspect that the catalog includes all the potential memories that can be experienced by an individual. Will painful memories some day be removed entirely and not merely covered up or repressed? Or, come to think of it, can pleasurable memories be inserted? We can live only one lifetime, but perhaps we can remember five or six.

It is also suspected that aging is a natural process allowed for by the system for ordering catalog parts. The hope arises that if we can interfere with that system we might perhaps make it possible for ourselves to live forever—or at least until we get ourselves accidentally smashed flat by a steamroller. Or suppose we start at the beginning—with a fertilized ovum. As soon after fertilization as possible we can read its catalog and decide whether to let development continue. Why bother if the catalog is seriously defective?

But what if one or two items are missing and the catalog is an unusually good one otherwise? Why not supply the missing items? In fact, why not experiment with changed items or with altogether new items? May we not, in that fashion, create new breeds or new species? If we work with men, do we create supermen? In short, do we become gods, creating men (for better or worse) in the image of our own ideals?

There is no telling how close we are to being able to wield this frighteningly divine power, but considering the rate at which we are progressing, we may be only decades away. It is possible that the man, who will first alter a fertilized ovum and bring about the birth of a frog (let us say) that is not quite the frog he started out to be, is alive today. The man is perhaps also alive today who will someday remove cells from an inadequate gland, alter them, cause the altered portion to grow into a sizable culture, and then restore them with their healing power to the original organism. And millions alive today may well live to see it done.

Well, then, let's see what is behind all this. Living creatures, from bacteria to sequoia trees and from amoebas to whales, consist of cells. A very small creature, such as an individual amoeba, consists of a single cell. A human being consists of some fifty trillion (50,000,000,000,000).

Each cell is a kind of self-contained chemical and physical unit, though the many that may exist in an individual organism work in well-oiled cooperation. If we can find out sufficient details of the cell machinery for each different kind of human cell, we would then be able to figure out, without much trouble, the functioning of the human being as a whole.

Within each cell, thousands of different chemical changes are constantly taking place. Atom groupings called molecules are constantly undergoing reorganization. Here a large molecule may break in two; there two small molecules may join together. Here an atom may be split off; there it may be attached. One molecule may pass through a thin partition; another may not. The intricacy of the pattern of change, even where the smallest drop of living matter is concerned, a drop too small to be seen with the unaided eye, is unimaginably great.

But what brings about these chemical changes? Left to themselves, most of these changes would take place quite slowly; in some cases, incredibly slowly. There are present in each cell, however, certain intricate protein molecules called "enzymes." Each has a highly specialized surface built-up of definite atom patterns, and on that surface, one and only one kind of chemical change can take place rapidly. A cell may have thousands of different enzymes and only those chemical changes will take place quickly which will be brought about by those enzymes.

The pattern of chemical change will depend on which enzymes are in a cell and on how much of each enzyme is present. A human brain cell differs in its function from a human liver cell because the two cells have a different enzyme pattern. A human liver cell differs from a chimpanzee liver cell because the two enzyme patterns are not quite the same.

Indeed, no two human beings (barring identical twins) have quite the same enzyme pattern in their respective cells, so that organs and tissues differ from person to person. This can be a matter of life and death for a man may need a new kidney and yet his body will reject the transplanted kidney of any other individual because the enzyme pattern of the new kidney will be alien. Only a kidney from an identical twin (if the patient is lucky enough to have one) can be sure of being acceptable.

Why do enzymes have this vast potentiality for difference among themselves? Well, each enzyme molecule is made up of a couple of hundred smaller units called amino acids. There are twenty different amino acids in enzymes and an individual enzyme molecule can be made up of any number of these, arranged in any order.

Suppose you wanted to build up "words" out of the twenty letters of the alphabet from A to T, and make each word one hundred-fifty letters long. Suppose that *any* combinations of letters is a word so that the combination qertioplkjhgfdsacbn mlkjhhgfdsasddghjklk jhg fds aqeqertetitioioplokikolki ghjgtgfrfer fdsqaqas dfghjklkjhgfdsacdfgcfcfbnhnjmklk mnbnmnmiojplkojkkkjhgfds is one of these words. How many other words could you make up, if each different arrangement (however slight the difference) was a different word. Don't work it out; I'll give you the answer. Write down a three and then, after it, write down one hundred ninety-five zeroes.

If you want to say the number in words, you can come close by saying three thousand trillion, trillion, trillion, trillion, trillion, trillion, trillion, trillion, trillion, trillion, trillion, trillion, trillion, trillion, trillion, trillion. —And if you allow for words that are shorter than 150 letters, you increase the total number enormously further still.

That gives you a rough idea of how many different enzymes might conceivably exist. Every enzyme in your body can be different; every enzyme in every organism that has ever lived on earth can be different; and even then the potential capacity for variation in the enzyme molecule will still not have been exhausted.

Yet each cell has certain enzymes and no other. A human liver cell has one set of enzymes and a frog liver cell contains another set, and each liver cell sticks to its own. Indeed, when a liver cell in your body divides to form two new liver cells, each daughter liver cell has the same set of enzymes the original cell had. What makes each new liver cell "know" how to form the correct enzymes out of all the trillions of trillions of trillions that are possible?

Apparently, each cell has a set of instructions that it can pass on to other cells, and it would be of extreme importance to find and understand these instructions.

Sixty years ago, it began to be apparent that these instructions, whatever their nature might be, were located in certain bodies called "chromosomes" which were, in turn, packed into the central nucleus of the cell. These chromosomes are most clearly evident (under proper chemical staining) at the time the cells divide, when they look like a tangled mass of stubby pieces of spaghetti. Every species of creature has some fixed number of chromosomes in each of its cells. Human beings have forty-six chromosomes per cell. Since chromosomes always exist in pairs, we might just as well say that human beings have twenty-three chromosome-pairs per cell.

Before the cell divides, each chromosome brings about the production of an exact replica of itself (a process called "replication") so that two identical sets exist temporarily. When cell division is complete, each daughter cell has received one of these sets. In other words, each cell, as it is formed, inherits a complete set of instructions. The set of instructions originally present in the single cell that makes up the fertilized ovum is spread by replication after replication after replication to every one of the fifty trillion cells of the human body.

When a sex cell is formed by the body (an egg by the woman or a sperm by the man) it receives only a half set of the chromosomes; one of each chromosome-pair. *Which* of each pair? That is a matter of sheer chance. Each sex cell can have chromosome 1a *or* 1b, chromosome 2a *or* 2b and so on. Any combinationof a's and b's is possible over all twenty-three pairs. The total number of different chromosome combinations that are possible in the sex cells of a given human being are therefore eight trillion (8,000,000,000,000), no two of which are exactly alike, since the individual chromosomes of each pair are never exactly alike.

When a sperm cell fertilizes an egg cell, the fertilized ovum that results has a complete set of chromosomes—twenty-three pairs. One of each pair is from the mother and one from the father, so that the child inherits equally from each—though exactly *what* it inherits from each is determined by chance. Since the sperm and egg can each possess any one of eight trillion chromosome patterns, the final product can be any of eight trillion times eight trillion or sixty-four trillion trillion (64,000,000,000,000,000,000,000,000) patterns. Nor is this the limit since chromosomes can, in the process of sex-cell formation undergo subtle changes that further enormously multiply the number of different patterns possible. It is no wonder brothers and sisters are not exactly alike. (Identical twins originate from a single fertilized ovum—that's a special case.) In fact, the odds are enormously against any human being ever being exactly like any other human being who ever existed.

But what in the chromosomes contains the instructions?

Chromosomes are made up of two types of substance. One of these is protein in nature, as the enzymes are. The other is quite different; it is something called "deoyxribonucleic acid," which is almost invariably referred to by the initials DNA. By the mid-1940s, it proved to be the DNA which was the key component of chromosomes, rather to the surprise of biochemists everywhere, for they were betting on the protein. The DNA molecule is even larger and more intricate than the average protein molecule. It is made up of a string of smaller units called "nucleotides" and these come in four varieties, which we can call A, B, C, and D.

It might seem that a molecule made up of four different units is insufficiently complex to guide the formation of another molecule made up of twenty different units—but the nucleotides don't work in isolation. The nucleotides making up the DNA chain work in combinations of three; each combination (called a "codon") standing for a particular amino acid. There are sixty-four possible combinations of three nucleotides where each can be any of four different varieties. If you try to write down all of them; AAA, AAB, AAC, AAD, ABA, ABB, and so on; you will find exactly sixty-four.

Since you have sixty-four codons for twenty amino acids, you have a little leeway. Two or three different, but closely related, codons can each stand for one particular amino acid. Thus, ABB, ACB and ADB, may all stand for the same amino acid. This gives the situation a little redundancy. The print in the set of instructions may be slightly blurred, so to speak, and still be readable. Some codons may even represent a kind of punctuation, signifying where to begin an amino-acid chain and where to end it. A stretch of codons sufficient to yield the complete amino-acid chain of an enzyme is called a "gene."

There arises a question as to how the instructions get from the DNA molecule to the enzymes, since the former are deep in the nucleus and the latter are formed outside the nucleus. The DNA, remaining safely tucked away (as befits a valuable catalog of instructions) sends messengers.

The cell is capable of forming another variety of nucleic acid called RNA. A molecule of RNA can be formed against the genes in the chromosomes, molding itself, so to speak, on the DNA design. The result is a molecule of "messenger RNA," which moves out from the nucleus to the place outside, where enzymes are formed.

Present in the neighborhood of the enzyme-forming sites are twenty different varieties of fairly small molecules known as "transfer RNA," each of which is double-ended. At one end, a particular variety of transfer-RNA can fit onto some particular codon on the nucleotide chain of the messenger RNA. At the other end, the particular variety of transfer RNA can fit onto some particular amino acid. A group of transfer RNA molecules line up along the messenger RNA and at the other ends of the group, a lineup of amino acids is automatically formed in a specific order ultimately dictated by the order of codons in the messenger RNA—and this was determined by the order of codons in the gene within the nucleus.

How is the structure of the various genes (the catalog I spoke of at the start of this article) maintained accurately from cell to cell, and from parents to children to grandchildren? Actually, each DNA molecule is not merely a chain of nucleotides but a double chain. Each strand of the double chain is the "negative" of the other. When the cells are in the process of separation, the double chain separates into strands and each strand brings about the formation of another that fits its own mold. Each forms its own "negative" and in place of one double chain, you have two double chains, each exactly like the original. This is the chemical level of replication, and the basis of what is observed optically in the replication of chromosomes.

Naturally, the process is not always perfect. Mistaken replication can take place; a wrong nucleotide can slip into line here or there, and then maintain itself in future replications. Over the years and generations, a particular gene can, in this way, give rise to a whole family of similar but not identical genes. The process by which a gene produces another not quite like itself is called "mutation." A mutated gene produces a mutated enzyme which may or may not work quite differently from the natural one. Often the mutated enzyme won't work at all and cells containing it are distorted in their functioning to that extent.

If the mutation takes place in a sex-cell, the offspring may be radically different from the parents. One missing enzyme can produce an albino, or a child suffering from a certain variety of mental deficiency. Occasionally, a mutated gene–enzyme individual may possess an ability that is particularly useful under certain conditions. The processes of natural selection allow those mutations that fit to flourish and procreate, while slowly eliminating those that do not. This combination of mutation and natural selection is the driving force in evolution.

But if the fertilized ovum begins with a particular set of genes and passes this on to all the cells that arise from it, don't all fifty have identical sets? The answer is, yes. But why, then, aren't all human cells alike? Why are some of them skin cells, some liver cells, some brain cells and so on?

That is because our gene catalog is not used to the full. Individual cells may have several thousand enzymes apiece, but there are enough DNA molecules in human chromosomes to direct the production of about five and a half million enzymes of a 150 amino acids apiece. In other words, a cell makes use of only about one thousandth of the catalog. Which thousandth? That depends on the cell. As it turns out, while there are some genes which act to bring about the construction of specific enzymes, there are others which act as controllers or regulators of the first group.

We could say that each working gene has somewhere near it a regulating gene that can turn the working gene on and off. This is clearly necessary. If a working gene were continually on tap, the cell would be flooded with some particular enzyme. For the cell to work properly, the enzyme is needed in proper supply; not too little but also not too much. To be sure, enzyme molecules are fragile and "wear out" so that they must be periodically replaced and for that the working gene must be capable of being turned on periodically throughout life. When the supply of a particular enzyme is adequate at some particular time, however, the working gene must be turned off.

How does the regulating gene turn the working gene on and off? Presumably, some substance is formed by the regulating gene, which coats the active codons of the working gene and prevents messenger RNA from being formed.

But what turns the regulating gene on and off? What tells the regulating gene when it is necessary to turn off the working gene? Well, the working gene produces an enzyme which, in turn, brings about a certain chemical reaction in the cell. When the enzyme is present in ample quantity, the chemical reaction proceeds at full speed and the substance produced by those reactions pile up. The presence of those substances in the cell act to turn on the regulating gene which, by its action, turns off the working gene. Now, as the enzyme in question is no longer formed and as wear-and-tear reduces its supply, the chemical reaction involved slows down and the substances produced by it fade off. In the absence of those substances, the regulating gene is turned off and that turns on the working gene. More enzyme is formed. The genes in the chromosomes are controlled by feedback, just as the thermostat in your home is.

In any given type of cell, it would seem that most of the genes are permanently turned off. The pattern of permanently turned-off genes differs from one type of cell to the next, and it is the nature of the pattern that decides the type of cell, in fact.

But we are entitled to ask what causes the pattern to shift from one type of cell to another, if we start with a single fertilized ovum with one set of genes. Are all the genes in a fertilized ovum turned on? If so, what turns some of them permanently off in one set of cells and others off in other sets, as the ovum divides and redivides? If the fertilized ovum already has a pattern of turned-off genes, on the other hand, what changes that pattern in different ways as the ovum divides and redivides?

Presumably, as the ovum divides and redivides, not all the daughter cells are subjected to the same environment. Some are formed out of portions of the ovum that are rich in food supply; others out of portions that are relatively poor. Some are formed near the point where the sperm happened to enter; some far from it. Some find themselves near the outside of the ball of cells into which the ovum is soon changed; some find themselves inside.

The cells that are subjected to one environment are affected in such a way as to turn certain genes on and others off. This will result in the production of specific chemical substances, which may leak out of the cell and influence neighboring cells in some way. One can picture an intricate pattern of one-domino-knocking-over-another in which small changes at the very start build up and up, producing additional small changes here and there, each of which produces still more changes, until finally a completed organism with differentiated structures is constructed.

Once an organism reaches full growth, certain limited pattern-changing phenomena may continue. It is possible that every sense-perceptive unit we take in (every sight, sound, smell, taste, touch) affects the enzyme pattern in specific brain cells in such a way that a memory system is set up. What we experience, we can later recall—or forget. There may also be an automatic process of pattern-change that proceeds slowly and inexorably over the years and brings on the regular changes of aging. In other words, the body automatically prepares for ultimate death even when no external agency forces it—for the death of the individual is essential to the life and evolution of the species.

All this is easy to say. We can talk about working genes, regulating genes, and substances that turn genes on and off. What, however, are the substances and exactly how do they work? Scientists don't know the details yet, to be sure, but if the basic structure is worked out, can mere details be more than a matter of time? And once we map out those details, we will not only have the catalog, but we will understand how it works. Perhaps we can then learn how to shut off the supply of some items and order others. Perhaps we can improve memory or even intelligence. Perhaps we can even manufacture memory and devise methods for providing men with fantasy lives—surely the ultimate in vicarious entertainment.

We might alter the progressive changes in gene pattern to slow down the aging process in individuals where this seems advisable, or stopping it altogether. We might shunt cells into completely new directions and guide evolution into paths that the hit-and-miss procedure of ordinary mutation might not stumble across for eons. Where the cells lack certain genes altogether, we might learn to insert them, so that a superior human being need not be lost for the lack of one trifling (but essential) cellular requirement. In short, the human being (and other organisms as well, mind you) could be reduced to machines whose workings we understand, regulate, and alter.

Whether we have the wisdom to do all this seems very dubious. Who can decide how safe it is to offer human beings fantasy lives? Who can decide which individuals deserve extended lifetimes; in which direction evolution is to be directed; what new varieties of human beings are to be formed? In short, who among us are gods?

There may be an answer. Perhaps by experimenting on a very limited scale, we may stumble across ways of producing human beings with far greater wisdom than ourselves, and to these, we may hand over the task of improving the human race as a whole. The thought is perhaps not a palatable one, but as we look about the world today, surely it must begin to appear that almost *any* risk is worth taking, for almost nothing we might do can possibly make matters worse than they will shortly become if we do nothing.

As the 1970s wore on, people working with genes learned to manipulate them in such a way as to offer the world the chance of redesigning species, or even creating new ones.

Now there grew a general fear, similar to those with which I ended the last two essays, but one that grew so strong that it threatened to go too far and to put an indiscriminate end to scientific research.

The following essay is, in part, an attempt to argue against the extremes of fear in this respect.

17

THE GENE SCENE

You are a mass of chemical reactions; thousands of them. Within you, electrons continually slip from atom to atom; atoms and groups of atoms slip from molecule to molecule; small molecules join into larger ones and large molecules split into smaller ones. All the different reactions are carefully orchestrated into a delicate balance capable of shifting one way or another with great precision.

These thousands of chemical reactions, all working together, changed you from the barely visible ovum that was your first existence into your adult self. They continue, now, enabling you to think and move, to grow and heal, to adjust to your environment in a thousand ways.

Your chemical reactions are not quite like those of your neighbor in every detail, which is why you are two different people. The chemical reactions in a giraffe, in an oyster, in an oak tree, are all significantly different from yours and from each other, which is why initial sparks of life, in developing, move unerringly toward maturity in the form of human beings, giraffes, oysters, or oak trees.

Each chemical reaction is governed by a complex molecule

called an "enzyme." A particular enzyme will bring about a particular rapid shift of electrons or atoms or groups of atoms—a shift that, in its absence, would take place only very slowly if at all. Each living thing on Earth has its supply of different enzymes. Depending on the exact number and nature and efficiency of its enzymes, a particular chemical reaction system is produced that directs the characteristics of that living thing and makes it what it is.

Each enzyme is produced according to the specification or blueprint set up by another complex molecule called "deoxyribonucleic acid" or, in abbreviation, DNA, and sometimes referred to as a "gene." DNA molecules are arranged in long chains within each living cell, forming what are called the "chromosomes."

Each individual DNA molecule is itself made up of a long chain of four different kinds of small units, which are capable of combining in any order. The number of possible different combinations is too vast to express easily; let's just say there are countless trillions of trillions of them.

Each different combination of the small units makes up a different DNA molecule; each different DNA molecule supervises the production of a different enzyme; each different enzyme can bring about a different chemical reaction. If two different enzymes bring about the same chemical reaction, one does so a little more efficiently than the other.

Each different set of DNA molecules within a particular organism (anywhere from dozens to thousands of molecules making up the set) produces a different chemical reaction system and therefore a different kind of living thing.

Each DNA molecule is capable of producing others exactly like itself ("replication"). When a single egg cell matures into an individual containing trillions of cells, the original DNA molecules have produced enough other molecules like themselves to supply all the cells. Individual living things also produce DNA molecules that can be passed on to offspring.

Each living thing is born with a collection of DNA molecules it has inherited from its parents. Since the DNA molecules are the duplicates of those of the parents, the chemical reaction systems that result insure that the young resemble the parents. Dogs have puppies and cats have kittens; never vice versa.

Yet if DNA molecules are capable of replication, the process is exceedingly intricate. Occasional mistakes take place and sometimes a DNA molecule may produce another that differs from itself in a minor way. This different DNA molecule then replicates and persists, though occasionally, it, too, produces a variant. Such spontaneous changes in DNA molecules are "mutations."

There is a continuing drizzle of mutations and that is one reason why children need not be *exactly* like their parents, or one brother like another brother. Evolutionary pressures take advantage of these accidental differences so that, over the eons, new species are formed and a vast collection of different kinds of living beings arise from what may have originally been a single bit of microscopic life drifting in the primordial ocean over three billion years ago.

Evolution has always worked blindly, depending on whatever mutations happened to occur and on whatever environmental conditions did to make some mutations more successful than others ("natural selection").

Human beings, however, can substitute intelligent direction for chance. Scientists can, in effect, create their own mutations, eventually design specific mutations, and then decide which mutations might be worth encouraging into continued existence.

Right now, it is bacteria that are the chief target for such "genetic engineering." They are simple, one-celled organisms, each very small and each multiplying rapidly, so that many mutations can be studied in a small space and over a short period of time. Furthermore, bacteria are chemically complex. They can carry out a number of chemical reactions that more complicated organisms cannot, so scientists have a particularly wide variety of different DNA molecules to work with.

The chief technique used by scientists right now is to split various DNA molecules into smaller parts and then recombine them. For that reason, the work is said to be done with "recombinant DNA."

Since the DNA parts can be put together in new combinations, the recombinant DNA can represent a completely different gene, and the bacterium that possesses such a gene will have chemical properties quite different from those of its unaffected relatives.

By the manipulation of DNA molecules, biologists can learn the intimate details of their workings, and what is learned from bacteria can be applied to more complex organisms, even to ourselves. All living things, after all, however simple or complex, depend on DNA molecules as chemical blueprints, and all do so in essentially the same way.

It should also be possible to use this kind of research for the production of new strains of bacteria with new and very useful chemical abilities.

Diabetes, for instance, is quite a common disease, and diabetics need insulin if they are to live normal lives. Insulin comes from the pancreas of mammals, and there is only one pancreas per slaughtered animal. That means insulin is in limited supply and the quantity available cannot be easily increased.

Insulin, however, is formed through the action of a particular gene. (It is because that particular gene is defective in one way or another in some human beings that those human beings develop diabetes.) What if an appropriate DNA molecule, capable of directing the formation of insulin, is deliberately formed out of various DNA fragments and a strain of bacteria is developed that contains the designed DNA molecule? Insulin could then be obtained from bacterial cultures in any reasonable amounts.

We might design bacteria capable of manufacturing other hormones; or of producing certain blood factors needed to clot blood, factors which hemophiliacs lack; or of forming new and highly specific antibiotics; or of putting together vaccines for use against those still smaller disease agents, the viruses.

Nor need we design bacteria only to be chemical factory workers, however. They could be farmers as well.

Nitrogen-containing compounds in the soil are essential to plant growth. As they are incorporated into plants and washed away by rain, the soil would gradually grow barren. Yet there are millions of tons of nitrogen gas in the atmosphere that plants can't use.

Some soil bacteria, however, have the unusual ability to combine free nitrogen in the air with other substances to form nitrogen-containing compounds and replace that lost from the soil. The activity of such "nitrogen fixing" bacteria is thus something that all higher forms of life depend upon. We might design bacteria that can fix nitrogen still more quickly and efficiently, and use them as a new kind of fertilizer that can make our agricultural yield double and redouble.

Then, too, we might design bacterial strains that can make

food by using the energy of sunlight as green plants do. Or we can design strains that will use sunlight to split the water molecule into hydrogen and oxygen, thus giving us a very convenient energy source that need never run out.

At the other end of the scale, bacteria are the great scavengers. Decay bacteria break down all fragments of matter that had once been living, even fragments no other kind of life can handle. If their work came to a halt, the world would become littered with undecayed scraps of indigestible matter and these would accumulate till all life stopped.

Those scavenging activities might be improved. Suppose bacteria are developed that are exceedingly efficient at turning wood and straw into sugar or at absorbing hydrocarbon molecules and converting them into sugar and protein. The hydrocarbon-absorbing bacteria could be used to mop up oil spills and greasy residues of all kinds—not only removing them from the environment, but converting them into material that, after a number of stages of eating and being eaten, could reach our own tables.

Bacteria might also be developed that can break down plastics which have been specially treated in preparation for discard. Or bacteria might be developed for the purpose of collecting and concentrating traces of useful metals from wastes or from seawater.

The possible advantages of genetic engineering are obvious. Are there any disadvantages?

Some people object to this interference with what they consider the natural course of evolution.

In doing so, however, they are about ten thousand years too late. Human beings have been tampering with evolution ever since they domesticated animals and developed agriculture.

Prehistoric human beings may have known nothing about genes, but they knew how to direct the mating of anima's and the fertilization of plants and how to select the offspring and the seedlings they wanted to keep.

The domestic animals and plants that we now have resemble only distantly the wild ancestral versions from which they were derived; they exist in numerous strains quite different from each other; some could not exist at all without constant human care. They are all the results of a primitive kind of hit-and-miss genetic engineering.

What's more, we have steadily encouraged the growth and expansion of our domestic animals and plants at the expense of wild species.

Now our greater abilities will help us do with germs what we have done with plants and animals; to develop domesticated and useful breeds at the expense of the wild.

Is it all bad? Have even agriculture and herding hurt the ecological balance of species? Probably, yes, but that is the price we have paid for each of us wanting to live longer, better, and more securely.

We can't give it all up. We can't "restore nature" by somehow returning to a food-gathering existence after the fashion of the chimpanzee. The earth could then only support about twenty million people. What would we do with the other four billion?

But we might move forward and, through genetic engineering, help improve the ecological balance of species. Domesticated bacteria that improve the fertility of the soil, remove wastes, and even perhaps supply food directly, might enable us to decrease our reliance on sheer expanse of land devoted to crops and herds (provided we also learn to limit our own numbers by reducing the birthrate). Continuing development of genetic engineering techniques might enable us to improve the ecological balance by prudent manipulation of the genes of the higher plants and animals.

—Yet what if, in developing our bacterial strains, there is a slipup of some kind? What if, unintentionally and unwittingly, a bacterium is developed that is harmful in some ways and escapes from the laboratory? What if it can produce a terribly, wildly infectious disease against which the human body has no defense?

It is this particular fear that is producing popular pressure against genetic engineering altogether, with proposals that all work in this direction be stopped.

Actually, few types of experiments in this field involve any danger, and even in those that do, the risk is quite small. Nevertheless, scientists in the field are aware of the risk, and in 1974 voluntarily stopped certain kinds of experiments until a proper assessment of risk could be made.

Now the National Institutes of Health have set up guidelines for such work. Some lines of research are forbidden altogether, even though the risks are small. Other lines of research, where the risks are even smaller, can be carried on in a few places that are thoroughly outfitted with such precautions as air locks and filters, special safety cabinets, clothing changes, and so on. Still safer grades of research are allowed lesser precautions, but even the most harmless requires careful procedures aimed at preventing trouble.

So far, no known damage has been done by research in recombinant DNA, and the fact is that if we wish to worry about the worst—infectious disease against which we have no defense and for which there is no treatment and which kills almost everyone it touches—it already exists *in nature*. Certain virus infections that have come to world attention in Africa lately, such as Lassa fever and Marburg virus, are highly infectious and almost 100 per cent fatal. They have been dealt with (so far safely) with the use of proper precautions, and further research in genetic engineering may teach us how to deal with such infections.

And, in the end, the riskier lines of research may be transferred to laboratories in space, orbiting the Earth and separated by thousands of miles of vacuum from the human population.

On the whole, the chances of benefits arising from research in genetic engineering are so great, and the chances of disaster are so small and so guarded against, that it would seem a tragedy to give up the former out of an exaggerated fear of the latter.

PART THREE

LIFE FUTURE

This is the fourth of a series of articles I did for a Chicago engineering firm. The first two were reprinted in my essay collection Science Past—Science Future *(Doubleday, 1975) under the titles "Technology and the Rise of Man" and "Technology and the Rise of the United States."*

The third appeared in my essay collection The Beginning and the End *under the title "Technology and Energy." In the afterword to that article, I said, "The remaining article which I entitled 'Technology and Communication' was bought and paid for and was supposed to be out in the summer of 1976. Despite several inquiries, however, I have seen nothing of this one. For that reason I can't include it here. If it finally does appear, however, it will go into some future collection."*

I received the published version at last in May, 1977, so here it is, as promised.

This essay, by the way, caused me some head scratching. It deals with the past of human endeavor, which led me to think that it ought to go into the book between "The Flaming God" and "Before Bacteria." However, it also moves into the future, and I finally decided that that was more important, and that it would be appropriate to place it later in the book. And that's what I did.

18

TECHNOLOGY AND COMMUNICATION

The ability to communicate is one of the hallmarks of being alive. Even the simplest creatures can alter their environment by the secretion of some chemical that will induce some appropriate response in another creature. A female moth, by releasing a small quantity of a particular substance, will be communicating the concept, "I am ready," and male moths will follow the scent from a distance of a mile or more.

The more complex a creature, the greater its ability to communicate messages in greater detail. Birds have various calls, mammals have various movements, sounds, and gestures, and all of it means something and is recognized by other members of the species or even by outsiders. When a skunk turns its back on you and raises its tail, or when a panther snarls and tenses its muscles, the wise person (or any other creature bright enough to recognize the sign) retreats at top speed.

A fixed set of sounds and gestures is, however, not enough to make an animal a human being. The chimpanzee can be taught dozens of distinct gestures but can carry on only limited conversations, dealing largely with its immediate physical wants and fears.

Even the greater ability of *Homo sapiens* to devise gestures falls short. Try teaching someone to do something as apparently simple as swinging a golf club with the proper stance, by dumb show only, and see how quickly you will lose your temper. It is the inefficiency of gesture, even when bolstered by elaborate signaling, that makes the game of charades so challenging.

As long as human beings were confined to gesturing as their sole means of communication, it is doubtful that even their greater brain would place them much higher than the chimpanzee in the development of social organization or technology.

With gesture alone, one human being cannot transfer any but the most primitive items of information to another. Each human being is condemned to work only with the ideas he himself can generate from a start provided by a simple social structure that takes care of his basic physical needs.

What is needed is some form of communication which is complex enough and versatile enough to be able to transfer abstract information from one human being to another, clearly and certainly. In human culture, universally, this form of communication has been speech. Information has been transferred by modulating sound in such a way as to make it possible for every idea to be represented economically and uniquely by a combination of phonemes.

The development of speech could not have come about until the brain had developed to the point where the speech center was sufficiently complex to allow the necessary delicate manipulation of lips, tongue, and palate that would control the sounds and make it possible to produce them rapidly. (The chimpanzee doesn't learn to speak because it simply can't speak; the speech centers in its brain aren't sufficiently advanced. The dolphin, with a brain as complex as ours, *can*, apparently, speak, but we don't know enough about its language to judge the nature or efficiency of its speech.)

Speech, for the first time, offered a form of communication, a method of transferring information, that could link a species in time as well as in space. With speech, those of an older generation could pass on their experience, their ideas, their painfully garnered wisdom to their children not only by demonstration, but by explanation. Not only facts, but deductions and abstractions, could be passed on. The new generation could begin with that and build upon it—so that a true technology could develop.

Where speech exists alone, however, the key to accurate transference of information over the generations lies in memory, and this has its limitations. Surprising amounts of information can be transferred over a period of time, especially if put into some verse form where rhythm and rhyme serve as memory aids—but the kind of dull statistics in which any complex society abounds cannot be.

By and large, then, we date the beginning of civilizations from the discovery of writing—a method for freezing speech and for making it possible to transfer unlimited amounts of information accurately.

The most complex society we know of that lacked writing was the Inca civilization of pre-Columbian Peru. The Incas did, however, possess a memory-aiding device, involving knotted ropes, which enabled them to keep records of statistical matters. With writing, the much more complex societies of the old world empires, such as Rome and China, were possible.

Writing itself is a rather tedious process, nevertheless, and the process of making duplicates is sufficiently time-and-energy consuming to make books short and their total number small—and sufficiently error prone to make them suspect after a time. Literacy in a writing civilization tended to be low since few people would, in any case, have access to reading material. Statistical records, even when thoroughgoing, rarely existed in many copies, and the destruction of a few temples might mean the loss of all written records of a particular society.

On writing by hand alone, the particularly great complexities of an industrial civilization would be difficult to maintain. To advance beyond simple writing, however, required a major technological advance.

This is not to say that technology had not already been involved in the matter of writing. Although the creation of symbols and (a matter of much more sophistication) an alphabet was a purely intellectual leap, turning that into something practical required the development of a writing surface and a writing instrument. A stylus could punch marks on a wet clay surface, a chisel incise them into stone, a brush could smear ink on papyrus or parchment—some alternatives were easier, some cheaper, some more permanent, but all were painstaking and slow.

Then in 1450, the German inventor, Johann Gutenberg developed the art of printing by movable type. The concept was a simple one (in hindsight), but Gutenberg had also to develop a proper metal alloy for casting type, one that melted easily and that expanded slightly on freezing, to produce sharp outlines. He had to devise proper techniques for keeping the type lined up accurately and for pressing it against paper evenly and firmly.

The development of paper was, of course, essential to a practical printing technology. It had first been made from bark, hemp, and rags in China about A.D. 100, and knowledge of the technique had slowly spread westward, reaching Germany about a century before Gutenberg's work. Nothing as cheap or as useful had been known to the ancients, and in time paper was to grow even cheaper and more plentiful once techniques for making it from wood pulp developed.

Printing caught on faster than any technological advance in history up to that time and it completely revolutionized human life.

It became possible to produce books in many copies, all exactly alike, very simply, quickly, and cheaply so that the reserve of knowledge and learning was increased enormously. It became possible for the literacy rate to increase, since there was ample material to supply the reading matter for the literate, no matter how many. It meant that at least some education was possible for everyone and not just for a very few.

With education and literacy spreading more widely through the population, a larger reservoir was formed, out of which competent scientists and technologists could arise. Furthermore, the thoughts and discoveries of scientists and technologists, once put into printed form, could quickly circulate through all of Europe so that to each scientist and technologist there were available not only his thoughts, but those of all his fellow workers in the intellectual vineyard.

Printing meant that, for the first time, a *community* of contemporary thought was possible and, as a result, scientific and technological advance shot forward at an increasing rate. It is surely no accident that the scientific revolution of the mid-sixteenth century began only after printing had established itself on the continent.

Technological advances in the art of printing improved Gutenberg's original invention in endless ways in later centuries, pointing always in the direction of greater and faster production of the printed word; of the wholesale outward-spreading of human thought at a continually growing rate.

The all-metal press was invented in Great Britain in 1795, and in 1844, the American inventor Richard Hoe produced the rotary press which made it possible to produce eight thousand copies per hour. In the 1880s, the German-American inventor Ottmar Merganthaler invented the Linotype which could automatically set and rectify an entire line of type. Meanwhile, the art of photography was invented and illustrations as well as print could be reproduced. In the last century the whole process of printing has been almost entirely automated.

On a more personal note, the American inventor Christopher Latham Sholes devised the first typewriter in 1867. A form of printing was brought into the home and office and, eventually, helped bring women out of the home and into the office.

The techniques of printing have reached the point now where millions of copies of particular books can be produced every year, of particular magazines every month, of particular newspapers every day. The flood of information has become large enough to overwhelm everyone, and no one can keep up with even a respectable fraction of what is available to know in even a narrow field—not even of what becomes newly available to know each day.

It is said that the quantity of scientific information generated in the laboratories, observatories, and studies of the world doubles each decade, so that the number of scientific papers published in this last decade is equal to the total number published in all the years preceding.

Surely this runaway inflation of information cannot continue forever, or even for very long, without breaking down the very process it should be serving. Science will be held back by lack of knowledge, not because that knowledge is not there, but because it is hopelessly lost in a vast snowdrift of other and irrelevant information.

If humanity is then to continue to advance in knowledge, technology, and (we can hope) wisdom, there must be new revolutions in the art of handling information. If the quantity of information has grown beyond the ability of the human brain to store it securely and retrieve it speedily, then some mechanical device for the purpose must be evolved.

The electronic computer is, probably, the answer. Devised just about five centuries after the printing press, the computer is very likely to revolutionize society in the same way the earlier advance did. What's more, the computer will surely do so far more rapidly, since the general acceleration of technological advance has brought the computer forward more each decade, than the printing press was each century.

The exact nature of the social changes that will be brought about by a turning-point advance in technology is not easy to foresee, but we can make some guesses. Thus:

An electronic computer is, essentially, an information-storage device which can, on demand, produce any particular piece of information or, on instructions, manipulate various pieces of information in its store and produce the results thereof.

We can imagine, in the end, a vast computerized library network, worldwide, into which the accumulated information of humanity can be poured and from which any portion of it can be retrieved at request.

It would not be just a matter of information in and information out. The computerized library might be asked to comb its vitals for the output of a particular writer or for all information of a particular sort on a particular subject. It might be asked for recent acquisitions of one kind or another, for recent comments on a certain topic, or advance, or report. In fact, a computer sufficiently advanced might be programmed to search through its own store of information, weigh and combine items from that store, and then deliver conclusions that would amount to advances of its own.

A partnership of man and machine could do more toward probing the deeper knowledge of the rules that govern the behavior of the Universe ("the laws of nature"), their uses and consequences, than either partner could separately.

Nor is it just abstract information concerning the Universe about us that might be fed into a computer. It can also be the repository of the highly personal and ever-changing information about each one of us. With the help of thorough computerization, the world may be forever in a state of instantaneous census taking so that the vastly complex statistics of its population can always be known and weighed in the course of the making of every governmental decision.

Computerization might, in this way, be the key to the first truly democratic system, since the weight of individual interests and needs and opinions could be estimated, with reasonable accuracy, for the first time. With each individual on record in detail, each individual becomes more a person and less a statistic, and to him *as* an individual, society can become more responsive.

Granted the existence of sufficient information stored in a global computer network, in a form capable of nearly instantaneous manipulation, we can also imagine a world without money—something that would represent one more step in a change that has been moving onward in the same direction through the history of civilization.

The last five thousand years have seen the increasing etherealization of financial transactions, and at an accelerating pace. To begin with, human beings bartered, exchanging material objects and services directly. Metal coins then came into use as a universal unit of barter. Paper bills, arbitrarily marked, proved a more convenient set of units than coins. Checks, which are personal bills of any denomination, were more convenient still, and credit cards lumped a month's checks into one.

Always, the direction was towards greater convenience, but always the demand was for a more complex society. Metal has intrinsic value, while paper is only as valuable as the economic stability of the society issuing it allows it to be. Checks imply a vast network of bookkeeping in the banking system, and credit cards demand computerization.

To continue in this direction, we might imagine ourselves placing all financial matters into the computer and allowing tiny electric currents, flashing this way and that, to do all that is necessary to accomplish what has always been (whatever its etherealization) a form of barter.

Suppose everyone's liquid assets (the amount to be used in financial transactions) are placed into the computer network and that each can be kept recorded in an appropriate device keyed to a thumbprint, or voiceprint, or chemical makeup of perspiration, or to something more subtle still. At some appropriate manipulation, a person could always learn the exact state of his assets.

Suppose that any transaction a person was involved in—the earning, depositing, investing, or spending of any sum of money under his or her asset level—was made final only when the devices of every party to the transaction were placed into a computer outlet which would then transfer the necessary sums, in electron pulses, from one card to another.

Taxation could become automatic as well. The government could automatically assign to itself a share of money out of every transaction, basing its cuts on the size of the deal and on the size of the asset level of the individual receiving the money. Other complexities would be taken care of and adjustments made (in a way more thoroughgoing, equitable, and personally fitting than is now possible), one way or another, at the end of each fiscal year.

The computerized handling of information might lend itself to abuses, yes. There's a chance of abuse in everything, and to wish for the advantages that a more complex society will bring us is to accept the added risk of greater opportunity of abuse. (A pauper need not fear having his or her jewels stolen, but most people would rather be well-off and accept the increasing risk of theft.)

At that, computerization, though it may seem to involve a loss of privacy and a threat of hidden manipulative graft and corruption, may supply the very techniques that would prevent abuse.

Every once in a while, one reads of computers doing incredibly stupid things or of being outwitted by some unscrupulous person—but that is always the result of improper programming and is a *human* shortcoming. As computers grow more advanced and complex, it is to be assumed they will be able to "learn" increasingly to recognize what might be improper programming and to question it. They will also become increasingly difficult to outwit.

As the possibility of tax evasion and financial irregularity becomes increasingly more difficult, people may give up trying, and honesty may become unavoidable and, therefore, fashionable.

There is the matter, too, of the speed of transfer of information, back and forth, among computers and people. Electronic communication has increased that speed of transfer enormously, but even the most advanced form of such communication currently in use is limited in capacity.

A television set can only receive so many channels from complex transmitting stations at a reasonable distance only. This means that our communication is quantity sensitive and distance sensitive, so that a television set must gear its output to the common wishes of millions and cannot serve you personally with all your quirks and quiddities.

What is needed is an electronic change analogous to that from writing to printing. Electronic communications must become so widespread and flexible that a kind of "electronic literacy" will be established, with every man owning and using his own electronic wavelength for transmission and reception, just as every man now can own and read (and even write) his own books.

A large step in this direction has become possible with the development of communications satellites which can serve as relays, receiving signals from one point on Earth's surface and sending it to another. As few as three such relays, properly placed, would suffice to blanket the Earth and make all points on its surface accessible to all other points.

Such satellite relays would make communication distance insensitive. Slight alterations in orientation could send signals to points thousands of miles from their origin, and it would be no more difficult or expensive to be in contact with the other side of the Earth as with the other side of town.

In 1965, "Early Bird," the first commercial communications satellite, weighing 88 pounds, was launched. It had the capacity to make available 240 voice circuits and one TV channel. In

1971, "Intelsat IV," weighing nearly a ton, was launched. It had the capacity for 6,000 voice circuits and twelve TV channels.

Now the Intelsat system has seven satellites in orbit and is used at 115 terminals on Earth in sixty-five nations. The full-time leased telephone traffic is 6,500 circuits and the world's investment in this system and others like it is over a billion dollars.

But this is only the beginning. As long as satellite communications are confined to the radio-wave spectrum, there is a sharp limit to the number of circuits and channels that can be set up. Communication would remain quantity sensitive.

The time may come, however, when laser beams will be used as communications devices—bare and unprotected in the vacuum of space but making use of fine optical fibers here on Earth. Since laser beams are made up of light waves, which are millions of times shorter than radio waves, there would be, correspondingly, millions of times as much room for separate channels and circuits.

With a virtually unlimited number of channels and circuits available, every person could have his portable phone, equipped for both sight and sound, and a circuit all his own. He could be in contact, whenever he chose, with any other person anywhere on Earth.

The printed word will be capable of being transmitted easily and widely, in a computerized space-relay world, so that each individual could receive facsimile mail transmitted from any point to any other point in a fraction of a second, facsimile documents, magazines, and newspapers, all produced, if he wishes, on the equivalent of his private-channeled television screen. And in the same way, an individual could, at any time, tap the centralized computer network for any book, any paper, any information.

As long as anything remains on the television screen as an electronic symbol, it is not permanent, of course, but must be wiped out if anyting else is to replace it. But then, not everything we look at need be in our permanent possession. To check the weather report, or the supermarket price list, or the latest baseball scores, is to be in possession of something ephemeral that needs only a glance. And anything which requires a more permanent status can be printed out and stored as a sheet (or collection of sheets) of material.

It would all amount to an extraordinary level of information democracy. Everyone in the world would have available to himself or herself the mass product of human thinking and gathered fact, and each can pick and choose that which amuses him or her, or that which he or she needs, for either a brief look or a permanent physical possession. —And that will go a long way toward making possible the growth of an evenly developed world geographically and socially.

The computer language can, of course, be translated into any ordinary language, and there is no reason why an individual cannot converse with a computer and get the information he needs in French, Bantu, Hebrew, or Cambodian. In fact, proper computerization may present the world with an almost instantaneous translating device that will make humanity, for the first time, of one tongue.

It may, however, prove convenient for mankind to adopt some common language in order to make the worldwide computer use more direct and simple. This is not to say that local languages will go out of use; rather that most people will become bilingual, speaking their native tongues among themselves and "Terran" to computers and the rest of the world.

To follow the line of least resistance, Terran might be the equivalent, nearly or even entirely, of English. Already, more people speak English as a first or second language, than they do any other language. Combine this with the fact that computers have been developed primarily in English-speaking countries, and English may prove a natural choice for Terran.

National prejudices and computer needs may, very likely, enforce modification. English, as it stands, is a gloriously flexible tongue, perfectly adapted to the needs of literature, but there are numerous changes that may make it a more efficient tool for the transfer of information via computer. The nature of the changes might be determined, in part, by a computerized analysis of the various languages of the world.

It may be, then, that the English-speaking people will be bilingual, too, speaking the idiomatic English of their literary heritage and the formalized, English-descended Terran of the computerized world.

Still another consequence of the computerization of information transfer may well be a fundamental alteration of the world's attitude toward education.

Throughout history the emphasis in education has been uniformity. One teacher teaches many students and in one large area all teachers may use the same detailed syllabus, and the students may be subjected to the same area-wide test system.

This means that little attention can be paid to individual differences in talents and aspirations among students. Any student too different from others in this respect is sure to be considered a failure, even where the differences involve great, but nonfitting, intelligence or talent.

With computers offering far more information than any teacher could, and with their output on tap on an individual basis, a computer-guided education could, however, be geared to adjust to the quality of each particular student. The nature of the subject studied, the rate at which and the depth to which it is studied, and even the manner in which it is studied, can be adjusted to suit individual student needs, desires, and personalities.

Furthermore, education may become a sufficiently well-adjusted process to fit not only young people, for whom it has traditionally been intended, but for people of any age. In fact, this broadening of the educational process may well become an absolute necessity.

As the death rate has decreased in the past century and as the birthrate has fallen less rapidly, the world's population has not only increased, but the age distribution has changed. A larger percentage of people are old. In 1900, for instance, 4 per cent of Americans were over 65; in 1970, the figure was nearly 10 per cent and by 2000 it is likely to be 12 per cent.

This gradual aging of the population may even accelerate if civilization is preserved into the twenty-first century, since pressure of population may force a drastic further decline of the birthrate and medical advance may further lower the death rate.

It will soon no longer be possible to keep only the young freshly educated, to place the burden of innovation and creativity on them primarily, while the mature adult is allowed to grow steadily more behind the times as he chokes on intellectual rust. There will not be enough young to maintain the deadweight of the many old.

The solution may be to make education a universal commodity, without regard for age. With the computerization of the educational process, there is no reason why human beings can't continue to study and learn what interests them for as long as they live. Indeed, the mere act of doing so will keep their brain more lively and functional, keep them more creative, and make them more apt to contribute to the world's happiness.

The education of the future will be geared, inevitably, to what we would now call leisure. Ever since the coming of the Industrial Revolution the meaningless muscular labor that turned most men and women into work animals has been done, to a greater and greater extent, by machines. People have therefore been able to work shorter hours in jobs that are, increasingly, administrative, intellectual, supervisory, service centered.

Much of the new work is, however, as meaningless, mentally, as preindustrial work was physically. Computerization and automation will lift the burden of repetitive mechanical sensing and fingering from the shoulders of humanity, and a good thing, too. Any job that can be done by a machine is beneath human dignity.

What will be left in a computerized world will be human curiosity and innovation. With the world run by machinery, human beings will be free to follow their interests freely. The leisure will not be that of today, however, where it is possible to watch television in a state of semicoma for six hours a day out of sheer lack of anything else to do. Instead, the availability of computerized education may introduce each individual to complex interests he might otherwise never have known he could have.

There will be many human beings who, as a matter of interest, will wish to engage in scientific research, in space exploration, in government, medicine, art, music, or literature—enough of them to help the computers run the world and make it a stimulating place in which to live. Others, on another level, may wish to be involved in entertainment, in sport, in stamp collecting, hiking, or playing chess. Some may even (though this is hard to conceive) be willing to do nothing but eat, sleep, and make love.

What's the difference? If society functions and if the individual is happy, who cares the exact route to happiness each individual takes, provided that route does not trample over the route of his neighbor?

And in all this, advancing technology, so often described as the route to "dehumanization," offers a new and deeper achievement of humanity.

It is only by advancing technology that we can hope to have

each individual able to tap the world's knowledge at will and to be educated at will in the field of his interest in his own way and at his own speed. It is only by advanced technology that we can make government truly democratic and capable of being concerned with each individual.

But if we turn our back on technology?

Even if we could do so without catastrophe (and we can't), it would mean marching backward into a world of mental and physical labor, in which people will have to be drudges, unenlightened by the gifts and products of human genius because they will have no time for them. Nor, except in very small groups, could people be dealt with individually since there would be no techniques to make that possible. Almost everyone on earth would be a faceless grubber doing work so subhuman that a mindless machine would do it better.

That would be the true dehumanization.

There is a general impression abroad that I am a creative person. I don't suggest that is wrong; I'm of the same impression myself.

Where the world and I part company, however, is at the point where the feeling arises that since I am creative, I must know the "secret" of creativity and can advise other people how to be creative. Others may think so, but I don't.

The fact is, I don't know where I get my ideas or how I come to put them together in written form. As far as I know, I do it only by thinking hard and working hard, and if there's anything less glamorous than that, tell me about it. That's not the kind of "secret" anyone wants.

Still, people keep asking me to do articles on creativity, and finally I decided to do a little thinking on the subject in order to see if I could come up with some notions. I did, and the following essay is the result.

19

ONE-TO-ONE

In 1856, James Abram Garfield (who, twenty-five years later, was to become the twentieth President of the United States and the second to be assassinated) graduated from Williams College.

On December 28, 1871, after having served in the Civil War and attained the rank of major general and, in the course of that war, having been elected to the House of Representatives, where he was now serving his fifth term, Garfield arose in New York to address a meeting of Williams College alumni.

At the time of Congressman Garfield's address, the educator Mark Hopkins was on the point of retiring, after having served for thirty-six years as president of Williams College. Hopkins had taught Garfield in the latter's senior year, and this is what Garfield had to say about the retiring president:

"I am not willing that this discussion should close without mention of the value of a true teacher. Give me a log hut, with only a simple bench, Mark Hopkins on one end and I on the other, and you may have all the buildings, apparatus, and libraries without him."

Was that the lachrymose and overstated comment of the old grad, remembering his college days through a haze of tears and distortion?

Perhaps! But even if it were, Garfield had hit upon something.

The quintessence of education is to have one student facing one teacher. If you are the student, let it be a teacher who knows your strengths and weaknesses, your interests and boredoms, your conventionalities and peculiarities; a teacher who can pay attention to *you* and who, most important of all, can learn from you and use that learning to teach you the more efficiently.

In that one-to-one case, you might, out of your own interests, out of the thinking and awarenesses of your own peculiar brain, go off in some new and unconventional direction in one respect or another, with the help and interest of the teacher, who would move cheerfully along with you. Since to do or think anything in a new and unconventional way is to be what we call creative, it therefore seems to me that the one-to-one relationship between teacher and student is ideal for the evocation of creativity.

The one-to-one case, however, has never been truly tried. At times and in societies where the education of children was the responsibility of the individual family, rather than of the society, those who were sufficiently well-off could afford to hire tutors. In those cases, the tutor was invariably of inferior social position and the student could despise and disregard him. This is certainly not the ideal.

Then, again, even in societies where one-to-one is not thought of in connection with education, an individual can, if he perseveres for a long enough time, reach the point where he is a postgraduate student under the guidance of a famous scholar on a nearly one-to-one basis. He is one survivor, however, among many who have fallen by the way, and he himself may have been scarred and dulled in the process.

But can a one-to-one basis of education be anything more than a dream, except for a very limited number of students who qualify through wealth, social position, or exceptional qualities of perseverance or survival? How can we expect there to be as many teachers as there are students?

In any society which requires at least some education for many or most of its members (regardless of whether the expense is borne by the society as a whole or by individual families), there must be mass education—one-to-many. To date, there has been no way out.

This can lead to certain results. A teacher can drill his many-headed class endlessly and most will learn to read and write with fair efficiency and will also learn to perform simple arithmetical computations and parrot back information at an elementary level in other traditional subjects.

The pressure, however, will be on uniformity because that is the only way the teacher can handle the class; and, what's more, it is the only way the students can handle each other.

It is as though a group of thirty people are being taken through a museum of vast extent in which the exhibits are of widely varying nature and are spread out over an endless number of rooms. The single guide has a limited amount of time to spend in the museum and has an itinerary he is supposed to follow, one that covers only a tiny fraction of all the exhibits.

It is clear that the guide can complete his task only if the thirty people follow dutifully in his footsteps, with none either moving ahead in impatience or lagging behind for closer examination. It would be worse yet if any member of the group, catching a glimpse of exhibits in a room the guide does not attempt to enter, leaves his fellows to dash into that room on his own. Any move that breaks the lockstep will enforce a delay while the errant member, or members, is brought back to the fold.

Under these circumstances, any member of the group who is constantly dashing ahead, or lagging behind, or disappearing in an unauthorized direction is disrupting the group. He is an endless irritation not only to the guide, but to the dutiful members of the group. All must resent the delays imposed by the actions of the maverick.

And so it is with the one-to-many situation of the classroom, then, which not only stultifies creativity—the move at a different speed or in a new direction—but which actually encourages the students themselves to suspect the very concept of creativity and to dislike and torment the creative person (and to continue to do so in later life).

This suspicion of the new and novel is not, of course, a product of mass education alone. All social interaction is a form of education and all mass activities of any kind can be disrupted by some individual who too consistently defies the consensus. In religion, in politics, in business, in the ordinary give-and-take of life, the maker-of-waves, the goer-against-the-current is invariably troublesome.

Still, mass education, because it grasps the individual at an early age, and because it carries with it a dense aura of authority and public approval, is a particularly efficient weapon against creativity.

There are people, to be sure, who remain creative despite all the pressures that mass education and society in general can bring against them. They do so, however, at considerable cost to themselves. It would be pleasant if life could be made easier for them if only because the course and advance of civilization is in the hands of the creative. They, in the end, make life worthwhile for us and it might be the decent thing to do to make life a little easier for them.

Well, then, is there anything we can do to encourage the creative society—one in which as many individuals as possible are creative and are punished for it as little as possible? (And we might as well realize that creativity is a very broad concept. We are used to thinking of it in connection with the fine arts, with literature, and with science, but one can be creative in any field. I have no trouble in conceiving of a creative stamp collector or a creative pole-vaulter.)

We might, for instance, reorganize our educational system so as to achieve the one-to-one relationship of teacher and student. We might allow every student, and not just a fortunate microminority, to have Mark Hopkins at one end of the bench and himself on the other.

But where can we find as many teachers as there are students? Isn't it a self-evident impossibility to have one teacher for every student?

This has been true in the past and remains true in the present, but it need not be true in the future. We might construct the teachers.

Suppose that our civilization endures into the twenty-first century (a supposition that can by no means be taken for granted) and that technology continues to advance.

Suppose that communications satellites become numerous and are far more versatile and sophisticated than those we've placed in space so far. Suppose that in place of radio waves, it is the incredibly capacious laser beam of visible light that is used to carry messages from Earth to satellite and back to Earth, while optical fibers, or still more advanced techniques, are used for communication on Earth itself.

Under these circumstances, there would be room for many millions of separate channels for voice and picture, and it can be easily imagined that every human being on Earth might have a particular television wavelength assigned to him, as now he might be assigned a particular telephone number.

We can then imagine that each child might have his own private outlet to which could be attached, at certain desirable periods of time, his personal teaching machine. It would be a far more versatile and interactive teaching machine than anything we could put together now, for computer technology will also have advanced in the interval.

We can reasonably hope that the teaching machine will be sufficiently intricate and flexible to be capable of modifying its own program (that is, "learning") as a result of the input of the student. In other words, the student will ask questions, make statements, answer test papers; and from all these questions, statements, and answers, the machine can gauge the student well enough to adjust the speed and intensity of its course of instruction and shift it, what's more, in whatever direction student interest displays itself.

Nor need we suppose that the teaching machine will be entirely self-contained or that it will be as finite as an object the size of a television set might be expected to be. We can imagine that the machine will have at its disposal any book, periodical, document, recording, or video cassette in a vast and thoroughly encoded planetary library. And if the machine has it, the student has it, too, either placed directly on a viewing screen or reproduced in print on paper for more leisurely study.

Of course, this is not to say that mass-education techniques will be, or can be, entirely replaced. There are subjects that require group interrelationships—as athletics, drama, and so on. There is also value, and even necessity, in the gaining of experience in human give-and-take. But there will also be the one-to-one where that is suitable.

To return to our earlier metaphor, the education of the future could be like a tour of a vast museum, during which. at stated times, the group can scatter, with each in the company of a personal guide who, for a time, is willing to take the individual through whatever rooms he may fancy.

And who will teach the teaching machines? It seems clear to me that the students who learn will also teach. If the student learns freely in those fields and activities that interest him, he will be anxious to demonstrate what he knows—all the more so if he has reason to think he has, through thought or experiment, added something not previously known or demonstrated in his chosen field.

All will be put back into the central hopper to serve as a new and higher starting point for those that come after. The pooled brain power and the encouraged creativity of the human species will race forward to heights and in directions it is impossible now to foresee.

But that is the future and we live in the present. Those of us who are adults have already lived through creativity-stifling education and what we have salvaged can probably stand improvement. What's more, we may have children who are in the process of passing through the grinder now and who may need help to survive as nearly intact as possible.

What does one do?

The first law of creativity is: *Thou shalt be interested.*

It might seem to you that other qualities are more important as starting points. For instance, to turn out my 198 books (this being the 198th), I have had to gather an enormous amount of information, remain at my typewriter persistently and long, and force my thoughts into careful and orderly English sentences.

Doesn't that take intelligence, perseverance, industry, drive, intuition, and a lot of other admired characteristics that people assume are inborn or come of a natural tendency? And if you don't have these things, aren't you ruined before you start?

No. All those words—intelligence, perseverance, industry, intuition—are on-and-off things that apply to some facets of a person's activities and not to others. A young man who, in school, may seem murkily dense and incurably lazy, with an attention span of zero and a comprehension index of less than that, might nevertheless prove to be a remarkable fellow when working under the hood of an automobile.

And as for myself, who can present certifiable evidence for all those good characteristics, I don't know which end of a hammer you hold when you are screwing a rivet. And I have neither intelligence, perseverance, industry, drive, intuition, or anything else for something that doesn't interest me, like shopping for clothes.

What it amounts to is that any normal person working in a field in which he is greatly interested is bound to be intelligent enough and persevering enough and industrious enough and intuitive enough to have a fair chance at proving creative. Have interest first, and all else will follow.

Well, then, how do you make a person interested?

You can't. Choose a subject which, for one reason or another, you think you ought to be interested in and try to force the interest. The chances are enormously on the side of failure. However, if you poke your nose into all sorts of things, you may, spontaneously and without pushing it, find something that interests you.

Suppose you find *nothing* that interests you. In that case, school damage has been serious indeed. Children are interested in almost everything. We have to spend half our time keeping them from being interested in trying to pick up flames and stepping out in the street to play with the speeding automobiles. With advancing age, some of the indiscriminateness is bound to fall off and interest focuses. It is not likely to dwindle to zero. If you're interested enough to read this book, that is evidence enough right there that it hasn't dwindled to zero.

And the corollary is that if you want your child to be creative, let him find his interest (assuming it is not at the level of playing tag with automobiles) and support him in it—even if it isn't *your* interest.

The second law of creativity is: *Thou shalt be courageous.*

To do or say something new and unconventional is a sure way of irritating those who, through their own inclinations or through inability to buck social pressure, never do anything but that which others do and that which has always been done. Unconventionality and novelty may even irritate some who are themselves creative, but in a widely different field.

The response to the irritation can vary in accordance with the permissiveness of the conventional society in which the creative individual happens to be embedded. At certain times and in certain places, the person who produces something new in thought or deed has been imprisoned, tortured, burned at the stake, or, in one well-known case, crucified. Fortunately, these are exceptional cases and, in the United States certainly, we need not fear extreme retaliation overmuch.

Nevertheless, milder consequences such as loss of jobs, social ostracism, or even mere ridicule can result in much hardship.

But then, who promises you a box of candy? If you want to be popular, accepted, welcome, one of the boys (or girls), you need only go along with the crowd and suppress any desire to be uncomfortably or too-noticeably individual. That's a high price to pay, though, as anyone who has tasted the joys of creativity would be willing to testify.

Some fields of creativity are more likely to draw ridicule and hostility than others; and some people are less courageous (stubborn?) by nature than others are. If your child has a field of interest likely to make trouble for him, you, at least, can supply an aura of encouragement for him that will give him at least some of the strength he will need to face the rest of the world.

The third law of creativity is: *Thou shalt be humble*.

Creativity is a double-edged sword. All of social and cultural advance has arisen out of the creativity of a minority of human beings. A great deal of evil which humankind has had to endure has also been the result of creativity. It would be nice if we could tell the difference between the useful and the harmful in our own creative thoughts and acts, so that we could save society the trouble of doing so—and in the process, condemn creativity and innovation as a whole for the harm that some of it does.

If you are a truly creative arsonist, for instance, who stands at the very top of your profession in skill and ingenuity, you will nevertheless be treated as a criminal if you are caught, and your creativity will be no excuse. For that matter, Hitler and Goebbels were creative to the point of genius in their ability to manipulate mass emotions, yet who would allow their creativity to serve as absolution for their crimes?

Then, too, while creativity in the arts always has its points of interest independently of the outside world, creativity in science must produce something that matches the Universe or it is worthless.

A creative astronomer may evolve tales of an erratically moving Venus out of poetic legends, and a creative archaeologist may invent long-past astronauts out of dimly understood or puzzling ancient artifacts—but if so, each is nevertheless entirely wrong, his work is useless, and not all the creativity in the world will, in itself, suffice to make him right.

—Which, of course, brings me to the point that insofar as anything in this article is novel and unconventional, it could be regarded as creative—but that the mere fact that my views may be creative does not mean that they are either correct or useful.

And I am aware of that.

For many years now I have been aware that I have been growing older, at least by the calendar.

Yet I don't feel any different inside my skull. There are some physical creaks here and there in the rest of my body, but I'm just as lively as ever, mentally.

You might point out that I can't really tell. If my mental processes are losing their flexibility and versatility, I would be judging that with the very brain that was deteriorating and would therefore be incapable of noticing it. If I were in the last stages of senility, and had only a brain in the last stages of senility to tell me so, what could I learn?

Fortunately, I have an objective way of telling. I write as much as ever and the material I turn out seems to be as good as ever; not in my own opinion only, but in those of my editors and readers who have no reason to lie to me.

How long I can keep this up, I don't know, but I'm hoping to keep it up as long as I live. Why should not a brain in constant use maintain its function with reasonable efficiency just as muscles do? If Swedish kings can play tennis into their late eighties, then I can write essays into mine, if I live long enough. —Which brings me to the topic of the following essay.

20

FAREWELL TO YOUTH

Suppose we survive!

There are many reasons for supposing that mankind and its civilization are facing heartbreaking crises in the immediate future, and that even if mankind survives, civilization may not.

But suppose, then, that not only we but also our civilization survives!

What would the conditions be that would make survival

possible? To answer that we have to consider the nature of the crisis that makes survival doubtful.

Primarily, it is a matter of population. The world's population stands now at almost 4 billion, far more than it ever has been before. That population, moreover, is increasing, at the moment, at a rate of 2 per cent a year, and that figure is also far greater than it has ever been before.

Combining the two figures, we see that there will be 80 million more people to feed next year than there are this year, and 80 million more the year after. This annual increase will itself increase each year as the population does. What's more, the great majority of these additional people will appear in the nonindustrial nations which can least afford to feed additional mouths. By the year 2000, unless disaster intervenes, the world population will be over 7 billion.

Disaster, however, is only too likely to intervene. It is already difficult to feed the 4 billion today. With energy shortages and worldwide inflation, it is hard to see how the food supply can be increased in the coming decades or even kept at its present level. Fertilizer and pesticides are becoming more expensive; energy for irrigation pumps and farm machinery has also become more expensive. Farm productivity is bound to decline in consequence—and the famines are already starting.

If, in the year 2000, we find ourselves able to look forward to a stable, working civilization for the twenty-first century, it will only be because we have beaten the population problem. There is no other alternative. Even if we imagine new sources of energy, new sources of food, new methods of food distribution, a new era of world peace, all developing marvelously in the next thirty years, that will only give mankind a short additional breathing space.

If, by the year 2000, the population is indeed 7 billion and is somehow well supported—and if it is *still* increasing at the rate of 2 per cent a year, that will mean 140 million additional people each year and 15 billion total population by 2040. How long will this small, man-ravaged planet stand up under this?

No, indeed, sooner or later the population problem must be beaten, or civilization will vanish under the weight of human misery, and the longer we allow man's ingenuity to postpone the date of decision the more difficult it will be to avoid the crash at the first misstep, and the more horrible that crash will be.

Once it comes, of course, the population will then decline

drastically as a result of a vastly increased death rate. The survivors may then never again have the ability to rebuild a technological civilization since the easy sources of energy will be largely destroyed, the planet's supply of metals will have been thinly scattered over the Earth, while much of the soil will have been ruined and may, perhaps, even have become partly radioactive as a result of nuclear war.

Therefore, if we are going to suppose that there will be a working civilization in the twenty-first century, we had better suppose that the population problem will be well toward a solution by the year 2000. By then, the people of the world will have agreed (with despair and ruin as the clearly visible alternative) to bring the population rise to a halt and even to reduce the population to some reasonable level—to perhaps no more than a billion people.

One way of bringing about such a population decline is to allow the death rate to rise to the point where it remains steadily higher than our presently still-high birthrate. Surely no sane man can look with approval at this. Who would wish to accept famine, disease, and violence as a way of reducing Earth's population unless he thought that would take place only in other parts of the world while he and his loved ones remained safe in some island of prosperity?

This cannot be, however. The Earth is now so interdependent, economically, that a vast disaster anywhere will touch all mankind. If half the world falls into ruin, the other half will be shipwrecked as well.

There remains the far more humane alternative of lowering the birthrate to where it remains steadily lower than our presently still-low death rate. If we assume a flourishing twenty-first century, then we must also assume that by the year 2000 the world will, one way or another, have begun to lower its birthrate and will intend to continue to lower it and keep it lowered for at least a century to come.

But if civilization survives, we can expect science and medicine to continue to win victories over disease and to continue to extend the average life span—forcing the birthrate lower still. And if that is the case, then we will have to look forward to a society utterly different from any that has ever existed on Earth before.

During almost all the history of the human race, mankind has lived under conditions where the death rate was high. Life expectancy varied from twenty-five to (very occasionally) thirty-five, so that where the birthrate and death rate were equal, half the population was under thirty. Where the birthrate was considerably higher than the death rate, so that the population was growing rapidly in numbers at the young end of the scale, half the population might be under fifteen. Through most of history, the number of people over forty was never more than perhaps 20 per cent of the whole, while the number of people over sixty-five was never more than perhaps 1 per cent of the whole.

In essence, then, almost all the societies mankind has experienced consisted largely of young people. Middle-aged people were a minority and old people were a rarity.

In a society in which the birthrate was lowered and kept low and in which life expectancy was seventy, then for the first time in the history of the human race, the accent would no longer be on youth.

If the birthrate and death rate were equal, half the population would be over seventy, and at least two-thirds would be over forty. And if the birthrate sank lower than the death rate, as would be necessary in the twenty-first century if civilization is to survive, then the percentage of the aging would increase even further.

As a matter of fact, we are having a foretaste of such conditions here in the United States now where, for a century, life expectancy has lengthened and the birthrate fallen.

In 1900, when the life expectancy in the United States was only about forty years, there were 3.1 million people over sixty-five out of a total population of 77 million, or just about 4 per cent. By 1940, there were 9.0 million people over sixty-five out of a total population of 134 million, or 6.7 per cent. In 1970, there were 20.2 million people over sixty-five out of a total population of 208 million, or nearly 10 per cent. By the year 2000 there may be as many as 29 million people over sixty-five out of an estimated 240 million, or 12 per cent.

If civilization is to survive, we must see this trend become worldwide. It will be farewell to youth and welcome to a world in which the accent will be on the middle-aged and elderly.

What would such a world of the middle-aged and elderly be like?

Many might at once suppose the following:

A world in which those over forty form a substantial majority would be one in which the spirit, adventure, and imagination of youth would dwindle and die under the stodgy conservatism and dullness of age. It would be one in which the burden of innovation and daring would rest on so few, and the dead weight of the old would add so heavy an additional burden that mankind would sink and fall apart. A world of the middle-aged and elderly, many would insist, would be a static and even a decaying world in which all that man has always valued would disappear.

But would this indeed be so? Are old people really a dead weight? Are they really a force for stagnation? The difficulty in answering this question rests in the fact that mankind has never experienced a really age-centered society.

In almost all societies Earth has seen—only our own excepted—old people were a rarity and were, for that very reason, valued. The occasional person who survived into old age could remember how things were before anyone else was born. He or she was the repository of past ways, the recordkeeper of tradition, the village reference book, library, and oracle.

But all those values, natural in a preindustrial culture, are gone now. Old people are too common to be revered for their longevity. Nor does anyone need their memories and knowledge of ancient ways now that we keep records on paper, on microfilm, and within computers.

In ancient times, it was the old men who ruled the church and the state. The word "priest" is from the Greek word for "old"; the word "senator" is from the Latin word for "old." But now, with old men too common to regard, we value youth and vigor in government, and politicians dye their hair and practice moving with an athletic swing.

In societies in which technology changed slowly, it was the old artisan, rich in experience and know-how, who could be depended on for the good job, the skilled eye, the shrewd judgment. Now technology changes rapidly, and it is the downy-cheeked college graduate we want, expecting him to bring with him the latest techniques. To make room for him, we forcibly retire the old at sixty-five or less, give each retiree a watch, a pension, and a ticket to a park bench.

In short, the notion that the middle-aged and elderly are necessarily a dead weight on society is a very modern notion, born of the fact that their numbers have increased and their functions have disappeared.

And yet what if this modern notion is right. What if it were the ancients who were wrong, valuing the aged only because there were so few of them and mistaking feebleness for wisdom. If that were so, then surely the outlook for the twenty-first century is dim since, if civilization is to survive at all, we will see then a society of the aging.

Let's think about it. Suppose we consider, in the first place, what would seem to all of us a self-evident example of the inferiority of age. Surely there can be no argument that old people are not as strong or as healthy as young people and are not as capable of hard and extended labor.

Since this is so, isn't it clear that the expanding population of the aging would contribute little to the world's work and demand much of the world's care, and that the contracting population of young would not be able to support it all?

Yet consider, on the other hand, that if society is flourishing in the twenty-first century, there will have been a continuing scientific and technological advance. The two-century-old shift of hard manual labor from the straining backs of mankind to the stolid wheels and levers of the machine will continue and become more extensive and intensive. The present trend toward automation and computerization will continue to broaden its effects, and the need for hard and extended physical labor will continue to diminish.

In the twenty-first century, the world's work will not be primarily a matter of muscle and sinew, and athleticism will not be required. The weakening bodies of men and women, as they grow old, will not decrease the amount of their contribution to the working of society by very much.

Then, too, we may also expect that medicine and its allied sciences will continue to advance, and we must remember that this implies more than a mere extension of the life span. We can see this clearly if we consider what has already happened.

People today live, on the average, twice as long as do our ancestors of a century-and-a-half ago—but that is not all. They are also healthier and stronger, on the average, at any given age than their ancestors were at that same age.

It was not just that people died young in the days before modern medicine. Even if they lived, the would have had to survive the battering of repeated bouts of infectious disease that we can now either prevent or easily cure; they had to live on damaging diets which were seriously deficient in vitamins and other nutritionally essential factors; they had no way of fighting diseased teeth and chronic infections, no way of ameliorating the effects of hormone malfunctioning, no way of countering dozens of other disabilities.

As a result, the aging of today are vigorous and "young" compared to those of identical years in the medieval days of knights and chivalry.

We can assume that this trend will continue into the future if civilization survives. Medical science, for sheer lack of other problems, is already beginning a wholesale attack on the disease of old age itself, and they may win some local victories over it. It may be that the aging people of the twenty-first century may be not very aging at all by our present standards.

Between the greater vigor of old men and women and the lesser demands on them, physically, in the next century, the whole concept of "youth" and "age" may be blurred and the growing percentage of the aged will simply not represent a physical drain on society.

Yet even if we dismiss the importance of the physical decay we usually associate with growing older, it remains possible to argue that an aging population threatens the breakdown of society in other ways. Consider:

As the need for brute muscle vanishes with increasing mechanization, and as the necessity for dull and repetitive mental labor diminishes with growing computerization, mankind will emerge into a world where the need will be precisely for that which is most fitting for our species' unique brain—creative and innovative thinking. It is that task that machines and computers will leave to human beings.

And might we not fear that this is precisely where age is most surely a liability. Is it not the experience of mankind that creativity and innovation are the hallmarks of youth? If we search through the history of human accomplishment, do we not find innumerable cases of young people achieving the new and startling and revolutionary against the clenched-jaw opposition of the old?

This is true in every field, even in science where, above all, the rule is that of constant change. Max Planck, who devised the watershed quantum theory that utterly revolutionized the science of physics, said that the only way to get a radically new theory accepted by science was to wait for all the old scientists to die. And most people would agree with that, even though there are many examples in history of men and women who, into advanced old age, remain creative and receptive of the new.

In that case, what can we expect of a world in which the old will increasingly outnumber the young? However young, strong, and vigorous the aging might be in a physical sense, what good will that be if they are nevertheless an intolerably static force? The very increase in physical well-being and longevity may merely serve to make the general stodginess of the aged linger on the longer and poison society the more thoroughly.

Will we witness a twenty-first-century society then, in which individuals are strong and vigorous but in which the whole is mentally immobile? Will we see a tiny minority of creative individuals who will not be numerous enough to prevent society from sinking into a deep slumber that will last till finally mankind dies of apathy and boredom?

But will that indeed happen? Is it utterly impossible to imagine a combination of age and creativity? Can one be old and yet eager to experiment and to experience the new?

Isn't it possible, after all, that we ourselves created the conservatism of age by assuming, to begin with, that it will exist? There are self-fulfilling prophecies, you know.

If people are told all their lives that it is a self-evident fact that with age they will cease being productive and creative, they will naturally believe it. They are then likely to settle down into stodginess because they have been preparing to do so for years. Would they do so to such a degree (or even at all) if they had always taken it for granted that someone who was creative and innovative in youth would continue to be so in age.

We have found examples of self-fulfilling prophecies in other groups. Children who come of a deprived environment at home or who suffer under the preconceptions of a teacher and who are expected to do poorly at school tend, indeed, to do poorly. When, for some reason, those same children are expected to do well by some other teacher, they *do* do well. Perhaps, then, if we calmly suppose that the aging will do well—

The contemporary American ideal is one of ethnic blindness and sex blindness. It is hoped that we will learn to let each individual fill that niche in society to which his taste and his capacities call him, without being hampered in that by any consideration of his or her ethnic origin or of his or her sex. Presumably, this will become a world ideal as well.

But we also need age blindness. A man should be able to do

the work he can do and wants to do without being hampered in this by any consideration of his age.

And if we are going to have age blindness, we need it, to begin with, in one all-important area. Throughout history, one vital social advantage has been reserved almost entirely for the young. This is the advantage of education.

Let's consider education for a moment.

In general, the average extent of education required of the young varies with social and financial position and with the economic nature of the society. The well-to-do can afford a longer schooling for their young than the poor can. Similarly, an industrialized society, with a great complexity of its parts, puts a certain premium on longer and more intensive education than is usually required by a nonindustrial society.

In the history of the United States, where there has been a steady rise in the standard of living and in the level of industrialization, the average length of education has steadily increased and the average age at which one has completed one's education has steadily risen.

Yet despite the gradual extension of the period of education, it continues to be associated with youth; it continues to have a kind of "cut-off date." There continues to be a strong feeling that there comes a time when an education is *completed* and that this time is not very far along in a person's lifetime.

In a sense, this lends an aura of disgrace to education. Most young people, who chafe under the discipline of enforced schooling and the discomforts of incompetent teaching, can't help but notice that grown-ups need not go to school. One of the rewards of adulthood, it must surely seem to the rebellious youngster, is that of casting off the educational shackle. To them the ideal of outgrowing childhood is to reach a state of never having to learn anything again.

The nature of education today, the ability to view it as the penalty of youth, puts a premium on failure. The youngster who drops out of school prematurely, and who abandons further education to take some sort of immediate job, appears to his peers to have graduated to adulthood. The adult, on the other hand, who attempts to learn something new, is often looked on with a vague amusement by many and is considered as somehow displaying a second childishness.

By equating education only with youth and by making it

socially difficult for the average person to learn after the days of formal schooling are over, we leave that average person with nothing more than the information and attitudes he gained in his teen-age years—and then we complain of the stodginess of age.

In a twenty-first century that will be clearly weighted in the direction of age, the most effective aspect of age blindness would be to break entirely with tradition and make education the right of all. There should be no feeling that education automatically stops at a certain age or a certain stage. It *may* stop for an individual if that is his free choice, but it may not; and those who freely choose to stop may later on as freely choose to return.

An age-blind world of universal education at all ages will make all the more sense if we remember that the twenty-first century will be a time of advanced computerization and automation. The work that must be done to keep society going will surely require the effort of only a minority of the population (that minority whose tastes and aptitudes will cause them freely to choose to do that work). What of the rest?

We can imagine an idyllic world in which the majority of the human race merely "enjoys itself," but enjoying one's self is hard work. A child who complains he has "nothing to do" is usually granted very little sympathy by parents with too much to do, but the child is in agony just the same. What if we have millions with "nothing to do"?

Education will have to be oriented toward leisure. As many people as possible must learn the various ways he or she might find as serving to fill one's life agreeably and pleasantly. Whatever it is you do interestingly and well will add to your own pleasure and to those of others; and if you learn whatever it is you learn—wood carving or computer designing or tennis—the greatest pleasure would lie in then teaching others, making use of who knows what new technical and psychological tools that will by then have entered the purview of mankind.

With teaching and learning the great tasks of life, the social pressure will be toward *continued* learning, and in an extended lifetime, it will seem natural to embark on a new field of knowledge or activity every decade or so.

The premium will be on learning new things throughout life and it seems quite reasonable to suppose that people, who expect from youth on to be learning new things all their life long, *will* learn new things all their life long.

In such a world the change of age pattern away from youth and toward more advanced years will *not* herald a decline in creativity and innovation. Quite the reverse, perhaps.

Is there still reason to worry, however, even if the individual's physical vigor and mental alertness remain high? What of the species as a whole even so?

Given an extended life span, a slow turnover of the generations, a declining population, will not the pace of human evolution slow down? Given also a stable environment, with humanity protected from unlooked-for change or trauma by caretaking machinery, will not the pace of human evolution stop altogether?

Might it not be that the human species will then stagnate? By devising a comfortable present, might it not be denying itself the potentialities of an advanced future? Might not the whole purpose of the species be lost in a quiet backwater of sameness through our excessive care for the safety and security of the individual?

Not necessarily, perhaps. Throughout all the long history of life, right down to the present, and for all species of life, including our own, evolution has progressed hit-and-miss. It has involved random changes in genes, blind slaughter of individuals, and the chance push of natural selection in this direction or that.

We can't quarrel with the process, for it works. At least it has produced man. It is, however, a process that takes *time*. From the first formation of a blob of matter that we might call alive, it took at least three billion years to form man.

Once man was formed, something utterly new came into existence: a brain complex enough to carry evolution into a new *purposeful* stage.

Mankind is like no other species, past or present. He has developed the capacity for using advanced biological engineering techniques. These will surely continue to be developed, and through them he may learn to control his own evolution.

One of the tasks of the twenty-first century will be to map out all human genes, determining their structure and their function both alone and in combination with other genes. It is a formidable task, and *all* the gene combinations cannot be tested in any conceivable time.

Advances can be made, however, and guideposts may be set up for the modification, replacement, recombination of genes.

The results can be tested in individual cells, in tissues, in organs, and eventually in intact organisms.

Without trying to pinpoint the actual biological techniques that will be evolved, we can nevertheless remain certain that humanity will be moving in some chosen direction, one slow step at a time—but millions of times more quickly than blind chance would do the job.

It seems a formidable and frightening task to us, and the thought of possible mistakes is a chilling one—but presumably the evolution engineers of the twenty-first century will be reasonably skillful at their job, will make fewer mistakes than we fear, and will be careful to make no change that is irrevocable.

Yet even if we assume that a program geared to education and leisure will prevent any decline in individual initiative and creativity, and that the development of genetic engineering will prevent any standstill of the species, it might be argued that all this may *still* not prevent the slow stagnation of mankind.

The world of the twenty-first century, as we are describing it, is a world without physical growth. The population cannot be allowed to grow; indeed it must decline. Nor can there be a continuing expansion in the use of resources since mankind will never forget the limited capacity of the Earth after the experiences with which the twentieth century will end.

There would, of course, be continued growth of knowledge and of the sophistication with which man's technology will be developed, but this is not an obvious factor, and it would, in any case, be devoted entirely toward making the Earth a comfortable home for a limited population. Man's whole drive would be toward establishing a stable policy of nongrowth.

Yet mankind has always lived with risk and adventure; it has been the possibility of failure, perhaps, that has made the striving for success so exciting. With success assured, toward what end will individual initiative be employed; and for what purpose of species improvement will genetic engineering be used? Might we not argue that with all the horizons gone, humanity will end in an eventual whimper anyway?

Yet horizons will *not* be gone. Already, mankind has left the Earth. On six separate occasions, two men have walked the Moon. In principle, a plurality of worlds has already opened up for mankind.

Under present conditions, the opening is not a practical one. As population increases, more and more of mankind's efforts must be turned to the bare task of survival. The small effort we have already made, just to reach the Moon and nothing more, strikes many people as representing an inordinate expense that might better have been put to earthly use. In the decades to come, when earthly needs will grow with fearful rapidity, the chance of finding the resources to support the exploration of space will dwindle equally rapidly.

In a world of low birthrate and declining population, however, in a world in which technology has survived and advanced, the outlook is entirely different.

In a world without war (for unless some way is found of preventing the incredible waste of energy and resources that even peacetime armies represent, our civilization will not survive) the exploration of space will serve as an emotional substitute. It will be *the* adventure, one in which all segments of mankind can share—a kind of war against the common enemy, the empty distances of the Universe.

With advanced techniques, space flight will be neither as expensive nor as dangerous as it is now. It will be possible not only to reach the Moon, but to engineer a living space under its crust that will be supplied, initially, with the necessary food, water, and machinery. Carrying on from there, with careful cycling and with the use of the Moon's crust itself for added material, the lunar colony will finally be able to exist independently of Earth.

The longer trip to Mars will surely be carried through next, and by the end of the twenty-first century, it may well be that there will be three human worlds, each markedly different from the other two.

The human initiative and creativity, which will still exist in a world of the aging, will find plenty of room for expression in the establishment, the expansion, and the improvement of the two new worlds. And those new worlds, in their early stages, may become new societies of youth.

The new techniques of genetic engineering will find their outlets in controlling and guiding the obvious needs for those changes in human anatomy and physiology that will better fit the new pioneers for life on the Moon and on Mars.

There is, in fact, no chance that mankind will suffer from the bordeom that comes with a world too secure and dull. It will take considerable time before the Moon and Mars are tamed to the point where the inhabitants of those worlds will find their lives boringly safe, and long before then, mankind will have moved still farther out.

Beyond Mars are the vast stretches of the outer Solar system, with worlds as impossibly huge as Jupiter, as middle-sized as the various satellites, and as small as the asteroids. How they can best be explored and exploited will be the task of the twenty-second century.

—And beyond the edge of the Solar system are the stars in gigantic numbers. Whether we can ever defeat the speed-of-light limit and make travel to the stars feasible in the ordinary sense, I don't know. Whether mankind can build huge ships that will serve as worlds in themselves and launch these outward with thousands aboard to spend generations in space, I can't say. Whether mankind, in expanding through the Universe, will encounter other intelligent beings who may either help or hinder, I have no way of guessing. But whatever happens, there is a horizon out there and while it exists, mankind need never be bored.

If, then, we can but get through the next few decades and survive the immediate crises that face us, there is the good possibility that we will thereafter enter a sunlit plateau of achievement and joy to which, in the foreseeable future, there will be no end.

I have an irrational fear of airplanes and therefore never fly. Sometimes I tell myself that the fear is not *irrational since one does read of airplane crashes. But then I lack any inhibitions about driving an automobile, even on holiday weekends. Mile for mile, I am told, the chances of harm coming to me in an automobile are greater than in an airplane, so that to discard the latter in favor of the former out of fear of harm is irrational.*

But then, it's the essence of irrationality that anything characterized by that adjective is beyond the reach of reason. I consequently expect to keep on driving and to continue to eschew flight.

When I was asked to forecast the future of transportation, I seized the opportunity eagerly. After all, if I considered what might happen to earthbound transportation, might I not find the airplane would be made obsolete?

The trouble was, though, that when I described some of the possibilities in better and faster land transportation, I had to admit the chances were that if these systems existed now, I wouldn't use them either.

Let's face it! Speed unnerves me. I'm the one in the taxi, who quavers, "I'm in no hurry, driver. You may slow down, please."

21

MOVING ABOUT

Through most of humankind's history, individual people have been relatively immobile. They have walked and run, to be sure, and hopped and jumped for that matter, but the distance they could travel in this way was small and they personally experienced only their native town and its immediate environs.

Soldiers could, and did, march thousands of miles even in ancient times, and horsemen or sailors could routinely make long trips. Long journeys took months, however, even years, and only a small minority participated. The average person remained in place.

It was only with the coming of the Industrial Revolution, with the steamship, the locomotive, and most of all the automobile and airplane, that the average person became able to call the whole planet his stamping ground; to move from point to point at will, whatever the distance, and in a matter of hours.

Now here we are, moving freely—except that we're choked by traffic jams and we're deafened by engine cacophony and we're paving over the land with an intricate network of highways and we're poisoning ourselves with the pollution of burning fuel and burning it, moreover, at such a rate that in thirty to fifty years there'll be none left to keep us moving and, finally, we're killing and maiming hundreds of thousands per year.

As we look into a future of steadily increasing population the world over, it might seem as though we must see more and more people moving about, more and more automobiles, buses and planes, larger and worse traffic jams, noise, pollution, waste, and death till it all crashes down into chaos.

If the population *does* go up steadily, the insatiable demands of more and more people for more and more food and services may indeed overload and destroy our technological civilization. If, however, the population boom is slowed, halted, and, finally, reversed, it may well be that no other problem will turn out to be insoluble and, in particular, that the future of transportation will be bright.

One answer to automobile pollution that is often advanced, for instance, is the use of electric cars, which would be relatively noiseless and nonpolluting. As long as we get our energy from fossil fuels, from coal and oil, electric cars are no solution, however. They only shift the pollution from one place to another. Instead of each car burning its own fuel, various electric plants will have to burn the fuel to produce the electricity. It may be convenient not to have various pollutants fill the air in the traffic-jammed cities, but the total amount of pollutants entering the atmosphere would be the same and, in the long run, the air will continue to get dirtier everywhere.

What is needed is some source of energy other than the fossil fuels which are, in any case, in dreadfully temporary supply. If we can learn to manufacture electricity from solar power, either by coating some of Earth's desert surface with solar cells, or by setting up a solar-power collecting station in space, then we would have a situation in which pollution would sink to a minimum.

Nor need we depend on solar energy as the only source. If we can develop controlled nuclear-fusion energy (not to be confused with the dangerous nuclear-*fission* energy we are now using), we would in this fashion have another almost unlimited electricity supply from a relatively pollution-free source.

We needn't expect electricity transport everywhere, however. Electricity-driven planes, for instance, are hard to imagine. There will always be times when liquid fuel would be convenient, but where will the liquid fuel of the future come from when the oil is gone? Some could be obtained from coal, but it would be more logical to use the cheap electricity produced from either solar power or fusion power.

Such electricity could be used to split water into hydrogen and oxygen. The hydrogen could then be combined with carbon dioxide to form simple alcohols and hydrocarbons. These are liquid fuels which, in engines, can combine with oxygen to yield energy as water and carbon dioxide is formed. You start with water and carbon dioxide and you end with water and carbon dioxide.

Furthermore, the fuel formed in this way will consist only of carbon, hydrogen, and an occasional oxygen atom. The pollution we suffer from today rests in impurities in coal and oil that contain other kinds of atoms: nitrogen and sulfur, for instance. We will thus have indefinite quantities of non-polluting fuel for the internal-combustion engines of the future, and our highways will see both fuel cars and electric cars.

If we solve the energy and the pollution problems, what about the matter of traffic jams and traffic accidents?

There we may look forward to the help of increasing automation. The traffic lights of today are inflexible. It is difficult to make them take account of different flows of traffic in different directions at different times. It should become possible in future years to have traffic lights that are capable of scanning the approaches to an intersection so that they can detect the relative density of traffic in the different directions and adjust the relative lengths of time that stop-and-go signals are given. In this way, maximum traffic flow would be permitted. In fact, whole networks of traffic lights might be computerized to react to traffic flow and to adjust themselves cooperatively in such a way as to let pulses of cars through, in this direction or that, in the most efficient manner.

Individual cars would use radar guidance that would make them capable of detecting obstructions at night, or in fog, and to detect deceleration in the automobile ahead. Of course, a driver might not notice or might disregard radar warnings, so it is likely that cars would be equipped with automatic devices that would slow or halt them at appropriate times. In this way, the highway carnage would drop considerably.

In fact, the automobiles of the future may well be altogether automatic. We can foresee one with a computerized "brain" that would retain the highway network fed into it for certain trips and that could then follow the most efficient route while guarding against obstructions and other cars and adjusting its speed to that of the traffic flow. It would require manual operation (or perhaps emergency resetting) only in case of unexpected detours.

The more convenient and safe we make the future automobile, however, the more we may expect people to use it. The traffic will remain heavy and all the computerizations possible won't keep it from being slow.

One solution will be to make use of transportation devices that can move more people in a given volume than automobiles can—buses or trains, for instance. Naturally, these will have to be improved, too, if they are to compete with the automobile. (Part of our present problem is that the mass-transport systems of today do *not* successfully compete with the automobile.)

The mass transport of the future may gain speed and comfort through the use of magnetic levitation. A central guide rail magnetized in fashion similar to the train itself will set up a repulsion (if the magnetic field intensities are great enough) that will lift the train a few inches above the rail. Since there would be virtually no friction, the train could reach a vibration-free speed of three hundred miles an hour.

An even more dramatic way of decreasing traffic jams is to get some of the traffic off the surface of the earth altogether and place it elsewhere.

There is at least a possibility, for instance, that the cities of the future will delve underground. That may seem strange to us, used as we are to open-air existence, but there are advantages to underground living.

The chief of these is the final defeat of weather, which is primarily a phenomenon of the atmosphere. Rain, snow, sleet, fog would not trouble the underground world. Even temperature variations are limited on the open surface and would not exist underground. Whether day or night, summer or winter, temperatures in the underground world would remain equable, and the only natural danger we would still have to fear would be the earthquake.

The defeat of weather is of prime importance to transportation since that would make popular the basic human device of walking. We all walk constantly, from here to there, from room to room, upstairs and downstairs. What keeps us then from walking in the open air? Chiefly, there is the objection that it is so often too hot or too cold or too wet or too windy for comfort. In an underground city in which there would always be nonweather, walking would be the rule on the horizontal, at least over short distances, and elevators for the vertical, just as it is in a large office building.

Over greater distance underground—from one end of a city to another, for instance—there might be automatically moving walks, like escalators on the level. There would be slow-moving "locals" with many places of entry and exit, and faster "express" strips for long-distance travel. (I described such an underground city in my book *The Caves of Steel,* published by Doubleday in 1953.)

It is possible we might have underground mass transit also. Trains may move through long tunnels by magnetic levitation, with the added refinement, more practical underground than on the surface, that the tunnels would be evacuated. With the absence of air pressure, trains could move very fast indeed, and transcontinental trips might be made at supersonic speeds, as fast as or faster than an airplane would do it and with much greater safety.

We can foresee a tunnel under the Bering Strait so that the major continents would be interconnected by such speeding vacuum transport from Capetown to Patagonia and all points in between. Australia would be the major populated area that would remain outside the main tunnel system.

* * *

The displacement of traffic could be in the other direction, too. It could be shifted above the earth's surface, as well as below it. Automobiles or other vehicles might ride on compressed-air jets. As in the case of magnetic levitation, friction would be reduced sharply, allowing greater speeds and less vibration.

The importance of jet transport is that it would broaden the possible routes of travel. Surface transport is confined to highways, subsurface trains to tunnels. A jet vehicle could move, however, wherever the surface beneath was reasonably flat, so that traffic density would drop everywhere except at relatively few bottlenecks, most of which could be engineered away.

There is an obvious disadvantage, to be sure, in the ability of a jet vehicle to move cross-country over private property and invading territory till now held secure from intrusion. It is to be taken for granted, however, that all technological changes involve social effects which will have to be taken into account by the lawmaking bodies of the world.

The most interesting thing about jet vehicles is, of course, that they would be able to travel over water about as easily as over land. In fact, they would go over water with greater freedom, since the water surface, except in gales and storms, is everywhere smooth and is not parceled out into privately owned parcels.

Since rivers could be crossed at any point, the traffic load on bridges will be decreased and they will be increasingly reserved for non-jet mass transport. Australia and other major islands, divorced from the continental vacuum-tunnel network could be reached by such jet vehicles.

Oceanic mass transport may be jet-assisted. Passenger liners and large freighters may travel on jets in good weather, making time; and may sink to the water surface when wind and weather make that necessary. Such jet-assisted water vehicles could cross land and would not have to bring their cargoes to the few ports that now exist, but to whatever portion of Earth's land surface would offer them the easiest and most convenient way of delivering their goods.

True air travel, at present, is almost entirely a matter of mass transportation, and giant planes can only take off and land at certain few and well-defined airports. That means that they contribute to the choking effect of traffic which must reach and leave those airports.

In the future, a new dimension of air travel could involve the use of the equivalent of air automobiles. This may come in the form of a Vertical Take-off and Landing (VTOL) plane. A VTOL plane would not require long runways but could take off in one backyard and land in another backyard. Undoubtedly, both take-off and landing spots would have to be specially designed to withstand the shock of departure and arrival, but any technological advance requires a host of fringe improvements. (The coming of the automobile meant the arrival of garages, paved roads, and service stations.)

VTOL planes could, in the end, be no larger than automobiles, hold no more of a load, be no more expensive, and be as suitable for individual use as automobiles are. And they, too, like the automobiles of the future, could be thoroughly computerized and automatic. Their advantage over ground vehicles is that the air is three-dimensional and roomier than the ground and that no roads would be necessary.

One can even imagine the coming of the equivalent of an air bicycle, in the form of a reaction engine strapped to an individual's back. One could fly then, personally, without the insulating effect of surrounding metal, and get the actual sensation of flight that one could not possibly get in any enclosing vehicle.

Individual jet propulsion might well become a great sport of the future (a kind of aerial scuba diving) and it might be valued by the young for the excitement of it, but it could only offer a minor contribution to actual transportation. It would be too slow and the unprotected human body is too fragile for such travel to be either economic or safe on a large scale.

Naturally, all these forms of travel I have mentioned require energy, whether for the purpose of turning wheels, pushing out jets, setting up a magnetic field, maintaining a vacuum or whatever. It all costs.

With solar power and fusion power, we should have all the energy we can reasonably need, of course, but that does not necessarily mean that we want to use all the energy we can, or have a mind to.

The use of energy always has its price. Direct solar power is often spoken of as nonpolluting, but that is not entirely true. If many square miles of desert are coated with solar cells, it can be argued that these would make use of sunlight that would have fallen on the Earth's surface anyway so that no pollution of any kind is added. The cells, however, may well absorb more light than the unprotected ground would. In this way, more heat would reach the Earth's surface than would have otherwise. This additional heat ("thermal pollution") serves to raise the Earth's average temperature very slightly.

If solar power is beamed to Earth from a space station, energy would reach Earth that would not have done so otherwise, and this would produce considerably more thermal pollution than surface-based solar cells would. Nuclear fusion would also produce thermal pollution (plus a little radiation which could probably be contained).

Thermal pollution can be dangerous. A slight elevation of the Earth's average temperature—not enough to disturb us unduly—might yet accelerate the melting of the ice caps and produce a disastrous rise in sea level. That may well limit our total use of energy until such time as we can learn to adjust the manner in which our planet maintains its temperature balance.

It would therefore be useful to cut down energy use by cutting down on transportation where that can be done. And a great deal of transportation *will* turn out to be unnecessary.

In many cases today, for instance, we transport mass not because we want the mass transferred from one place to another, but because we want the *information* carried by the mass to be transferred. Yet information has no mass and can be sent at only a small fraction of the energy expenditure that mass requires.

Thus, we are likely to send a document on a plane, and a person to hold the document, too, when all we need do is send the information on the document by electronic means at the speed of light.

In the future, our ability to transfer electronically will undoubtedly be greatly expanded. The time will come when communications satellites of advanced design are in the sky; when these are connected to each other and to the earth by laser beams of visible light which can carry a million times the number of messages than radio waves can; when earthly communications are connected by laser light passing through hairlike optical fibers. When that time comes, everyone can have their own television wavelength.

Everyone could receive documents, either on the screen, or reproduced in print. Offices and factories (which will be increasingly automated and computerized) can be monitored and their workings adjusted by television and telemetering. Business meetings can be carried on by way of televised images of the participants in three dimensions, images that will be indistinguishable from the real thing, as far as sight and sound are concerned.

Once fast-transferred signals can so easily substitute for slow-transferred mass, a vast number of business trips will not be taken, a vast amount of commuting will be skipped.

With that portion of transportation devoted to business trips dwindling and decaying, there will be an enormous saving of energy and much more room and comfort left for those who are traveling for social visits, for vacations, for fun.

If, then, humanity solves its population problem and learns to keep regional rivalries from collapsing into war and terrorism; and if it thus preserves its civilization and technological advance, then the transportation of the future may involve less energy, far less pollution and noise, and far more variety and pleasure than anything we have seen in the past.

Science fiction, which was my original stock in trade, routinely involves travel at any velocity, however high. After all, a speed of many millions of miles per second is necessary if there is to be travel between the stars with at least a reasonable fraction of the ease of, say, travel between the Earth and the Moon.

Yet, according to well-established theory and *observation, there is no way in which anything with mass (ourselves, for instance, and our vehicles) can transcend, or even equal, the speed of light in a vacuum—and that is a mere 186,282.4 miles per second.*

Most nonscientists find this limitation paradoxical and even abhorrent and seem sure that it is a mistake. Some way, they feel sure, exists whereby this limitation can be transcended.

The questioning is so persistent that I am frequently moved to discuss the subject and the following essay is my most recent attempt to reconcile public opinion with the way things are.

For myself, by the way, there is nothing upsetting about this speed limitation in the real Universe. As I said earlier, speed unnerves me. The speed of light, slow though it might be on the scale of the Universe, is quite fast enough for me. A little too fast, even.

22

THE ULTIMATE SPEED LIMIT

If you push something hard enough, it begins to move. If you continue to push it while it moves, it accelerates; that is, it keeps moving faster.

Why should there be a limit to how fast it can move? If we just keep on pushing and pushing and it will go faster and faster and faster and— Or won't it?

When something moves, it has "kinetic energy." The quantity

of kinetic energy possessed by a moving object depends upon its velocity and upon its mass. Velocity is a straightforward property, which is easy to grasp. To be told that something is moving at a high velocity or a low velocity brings a clear picture to mind.

Mass is a little more subtle. Mass is related to the ease with which an object can be accelerated. Suppose you have two baseballs, one of which is conventional and the other an exact imitation in solid steel. It would take much more of an effort to accelerate the steel ball to a particular speed by throwing it, than it would to make the ordinary baseball do so. The steel ball, therefore, has more mass.

Gravitational pull depends on mass as well. The steel ball is attracted more strongly to the Earth than the baseball is, because the steel ball has more mass. In general, then, on the surface of the Earth, a more massive object is heavier than a less massive one. In fact, it is common (but not really correct) to say "heavier" and "lighter" when we mean "more massive" and "less massive."

Well, then, back to our moving object, which has a kinetic energy that depends both on velocity and on mass.

If that moving object is made to move more quickly, by means of that push we're talking about, then its kinetic energy increases. This increase is reflected both in an increase in velocity and in an increase in mass, the two factors on which the kinetic energy depends.

At low velocities, the ordinary velocities in the world about us, most of the increase in kinetic energy goes into increase in velocity and very little into increase in mass. In fact, the increase in mass is so tiny at ordinary velocities that it could not be measured. It was assumed, for that reason, that when an object gained kinetic energy, *only* the velocity increased, while the mass remained unchanged.

As a result, mass was often incorrectly defined as simply the quantity of matter in any object—something which obviously couldn't change with velocity.

In the 1890s, however, theoretical reasons arose for considering the possibility that mass increased as velocity did. Then, in 1905, Albert Einstein, in his Special Theory of Relativity, explained the matter exactly, presenting an equation which described just how mass increased as velocity increased.

Using the equation, you can calculate that an object with a

mass of 1 kilogram when it is at rest, has a mass of 1.005 kilograms when it is moving at 30,000 kilometers a second. (A velocity of 30,000 kilometers per second is far greater than any velocity measured prior to the twentieth century, and even then, the increase in mass is only half of 1 per cent. It's no surprise that the mass increase was never suspected before this last century.)

As velocity continues to increase, the mass begins to increase more rapidly. At 150,000 kilometers per second, the object which has a "rest-mass" of 1 kilogram has a mass of 1.15 kilograms. At 270,000 kilometers per second, the mass has risen to 2.29 kilograms.

As the mass increases, the difficulty of further accelerating the object and making it move still faster also increases. (That's the definition of mass.) A push of a given size becomes less and less effective as a way of increasing the object's velocity and more and more effective as a way of increasing its mass.

By the time velocity has increased to 299,000 kilometers per second, almost all the energy gained by an object through further pushes goes into an increase in mass and very little goes into an increase in velocity. (This is just the opposite of the situation at very low velocities.)

As we approach a velocity of 299,792.5 kilometers per second, just about *all* the extra energy derived from a push goes into additional mass and just about *none* goes into additional velocity. If a velocity of 299,792.5 kilometers per second could actually be reached, the mass of any moving object with a rest mass greater than zero would be infinite. No push, however great, could then make it move faster.

As it happens, 299,792.5 kilometers per second is the speed of light, so what Einstein's Special Theory of Relativity tells us is that it is impossible for any object with mass to be accelerated to speeds equal to or greater than the speed of light. The speed of light (in a vacuum) is the absolute speed limit for objects with mass, objects such as ourselves and our spaceships.

Nor is this just theory. Velocities at very nearly the speed of light have been measured since the Special Theory was announced and the increase in mass we found to be exactly as predicted. Special Theory has predicted all sorts of phenomena which have been observed with great accuracy, and there seems no reason to doubt the theory or to doubt the fact that the speed of light is the speed limit for all objects with mass.

Let's get more fundamental. All objects with mass are made

up of combinations of subatomic particles that themselves possess mass; as, for instance, the proton, the electron and the neutron. These, and others like them, must always move at speeds less than that of light. All these mass-possessing particles have therefore been lumped together as "tardyons," a name invented by Olexa-Myron Bilaniuk and his coworkers.

There are particles, however, which at rest would have no mass at all (a "rest mass of zero"). These particles, however, are never at rest, so the value of the rest mass must be determined indirectly and not by direct measurement at rest. Bilaniuk therefore suggested the term "proper mass" be used to replace rest mass in order to avoid speaking of the rest mass of something that is never at rest.

It turns out that any particle with a proper mass of zero *must* travel at the speed of 299,792.5 kilometers per second, neither more or less. Light is made up of photons, particles which have a proper mass of zero, and that is why light moves at 299,792.5 kilometers per second and why that is known as the "speed of light."

Other particles with proper masses of zero, such as neutrinos and gravitons, also travel at the speed of light. Bilaniuk suggested that all such zero-mass particles be termed "luxons" from a Latin word for "light."

The celestial speed limit, that of light, has been a particular annoyance to science-fiction writers because it has seriously limited the scope of their stories. The nearest star, Alpha Centauri, is 25 trillion miles away. At the speed of light, it would take 4.3 years (Earth time) to go from Earth to Alpha Centauri, and another 4.3 years to come back.

Special Relativity's speed limit means, therefore, that a minimum of 8.6 years must pass on Earth before anything can make a round trip to even the nearest star. A minimum of 600 years must pass before anything can get to the Pole Star and back. A minimum of 150,000 years must pass before anything can get to the other end of the Galaxy and back. A minimum of 5 million years must pass before anything can get to the Andromeda Galaxy and back.

Taking these *minimum* time lapses into account (and remembering that the actual time lapse would be very much larger under any reasonable conditions) would make any science-fiction story involving interstellar travel extraordinarily complicated. Science-fiction writers who wished to avoid these complications would find themselves confined to the Solar system only.

What can be done?

To begin with, science-fiction writers might ignore the whole thing and pretend there is no limit. That, however, is not real science fiction. It is just fairy tales.

On the other hand, science-fiction writers can sigh and accept the speed limit with all its complications. L. Sprague de Camp did so routinely and Poul Anderson recently wrote a novel, *Tau Zero,* that accepted the limit in a very fruitful manner.

Finally, science-fiction writers might find some more-or-less plausible way of getting around the speed limit.

Thus, Edward E. Smith, in his series of intergalactic romances, assumed some device for reducing the inertia of any object to zero. With zero inertia, any push can produce infinite acceleration, and Smith reasoned that any velocity up to the infinite would therefore become possible.

Of course, there is no known way of reducing inertia to zero. Even if there were a way, inertia is completely equivalent to mass and to reduce inertia to zero is to reduce mass to zero. Particles without mass *can* be accelerated with infinite ease, but only to the speed of light. Smith's zero-inertia drive would make possible travel *at* the speed of light but not *faster* than light.

A much more common science-fictional device is to imagine an object leaving our universe altogether.

To see what this means, let's consider a simple analogy. Suppose that a person must struggle along on foot across very difficult country—mountainous and full of cliffs, declivities, torrential rivers, and so on. He might well argue that it was completely impossible to travel more than two miles a day. If he has so long concentrated on surface travel as to consider that the only form of progress conceivable, he might well come to imagine the speed limit of two miles a day to represent a natural law and an ultimate limit under all circumstances.

But what if he travels through the air—not necessarily in a powered device such as a jet plane or rocket—but in something as simple as a balloon? He can then easily cover two miles in an hour or less, regardless of how broken and difficult the ground beneath him is. In getting into a balloon, he moved outside the "universe" to which his fancied ultimate, speed limit applied. Or, speaking in dimensions, he derived a speed limit for two-dimensional travel along a surface, but it did not apply to travel in three dimensions by the way of a balloon.

Similarly, the Einsteinian limit might be conceived as applying only to our own space. In that case might we move into something beyond space, as our balloonist moved into something beyond surface? In the region beyond space, or "hyperspace," there might be no speed limit at all. You could move at any velocity, however enormous, by the proper application of energy and then, after a time lapse of a few seconds, perhaps, reenter ordinary space at some point which would have required two centuries of travel in the ordinary fashion.

Hyperspace, expressly stated or quietly assumed, has therefore become part of the stock-in-trade of science-fiction writers for several decades now.

Few, if any, science-fiction writers supposed hyperspace and faster-than-light travel to be anything more than convenient fictions which made it simpler to develop the intricacies and pathways of plots on a galactic and supergalactic scale. Yet, surprisingly enough, science seemed to come to their rescue. What science-fiction writers were groping toward by means of pure imagination was something that, in a way, seemed to have justification after all in Special Relativity.

Suppose we imagine an object with a rest mass of 1 kilogram going at 425,000 kilometers per second. This is nearly half again as fast as the speed of light, so we might dismiss it as impossible, but for a moment, let's not. We can use Einstein's equations to calculate what its mass would be at 425,000 kilometers per second and it turns out that the mass would be equal to $\sqrt{-1}$ kilograms. The expression $\sqrt{-1}$ is what mathematicians call an "imaginary number." Such numbers are not really imaginary and have important uses, but they are not the kind of numbers ordinarily considered appropriate for measuring mass. The general feeling would be to consider an imaginary mass as "absurd" and let it go at that.

In 1962, however, Bilaniuk and his coworkers decided to check into the matter of imaginary mass and see if it might be given a meaning. Perhaps an imaginary mass merely implied a set of properties that were different from those possessed by objects with ordinary mass.

For instance, an object with ordinary mass, speeded up when

it was pushed and slowed down when it made its way through a resisting medium. What if an object with imaginary mass slowed down when it was pushed and speeded up when it made its way through a resisting medium.

An object with ordinary mass had more energy the faster it went. What if an object with imaginary mass had *less* energy the faster it went.

Once such concepts were introduced, Bilaniuk and the others were able to show that objects with imaginary mass, going faster than the speed of light, did *not* violate Einstein's Special Theory of Relativity. (In 1967, Gerald Feinberg, in discussing these faster-than-light particles, called them "tachyons," from a Greek word meaning "speed.")

These faster-than-light tachyons have their own limitations, though. As they gain energy through being pushed, they slow down, and as they move slower and slower, it becomes more and more difficult to make them move still slower. When they approach a speed as little as 299,792.5 kilometers per second, they cannot be made to go any more slowly.

There are, then, three classes of particles:

1. Tardyons, with a proper mass greater than zero, which can move at any velocity *less* than the speed of light but can never move at the speed of light or faster.

2. Luxons, with a proper mass of zero, that can move *only* at the speed of light.

3. Tachyons, with a proper mass that is imaginary, which can move at any velocity *greater* than the speed of light but can never move at the speed of light or slower.

Granted that tachyons can exist without violating Special Relativity, do they *actually* exist?

It is a common rule in theoretical physics, and once accepted by many physicists, that anything that is not forbidden by the basic laws of nature *must* take place. If the tachyons are not forbidden, then they must exist—but can we detect them?

In theory, there is a way of doing so. When a tachyon passes through a vacuum at more than the speed of light (as it must), it leaves a flash of light trailing behind itself. If this light were detected, one could, from its properties, identify and characterize the tachyon that has passed. Unfortunately, a tachyon moving at more than the speed of light remains in a particular vicinity (say, in the neighborhood of a detecting device) for only an incredibly small fraction of a second. The chances of detecting a tachyon are therefore likewise incredibly small and none have as yet been detected. (But that doesn't prove they don't exist.)

It is perfectly possible to convert a particle from one class to another. An electron and a positron, which are tardyons, can combine to form gamma rays which are composed of luxons. A gamma ray can be converted back into an electron and positron.

There would seem, then, to be no theoretical objection to the conversion of tachyons to luxons and back again; or, for that matter, to the conversion of tardyons to tachyons and back again if the proper procedure could be found.

Suppose, then, that it were possible to convert all the tardyons in a spaceship, together with its contents, both animate and inanimate, into equivalent tachyons. The tachyon spaceship, with no perceptible interval of acceleration, would be moving at perhaps a thousand times the speed of light and would get to the neighborhood of Alpha Centauri in a little over a day. There it would be reconverted into tardyons.

It must be admitted that this is a lot harder to do than to say. How does one convert tardyons to tachyons, while maintaining all the intricate interrelationships between the tardyons, say, in a human body? How does one control the exact speed and direction of travel of the tachyons? How does one convert the tachyons back into tardyons with such precision that everything is returned exactly to the original without even disturbing that delicate phenomenon called life?

But suppose it could be done. In that case, going to the distant stars and galaxies by way of the tachyon universe, would be exactly equivalent to the science-fictional dream of making the trip by way of hyperspace.

Is the speed limit lifted then? Is the Universe, in theory at least, at our feet?

Maybe not. In an article I wrote in 1969, I suggested that the two universes that are separated by the "luxon wall," ours of the tardyons and the other of the tachyons, represented a suspicious asymmetry. The laws of nature were basically symmetrical, it seemed to me, and to imagine speeds less than light on one side of the wall and speeds greater than light on the other wasn't right.

Properly speaking, I suggested (without any mathematical analysis at all and arguing entirely from intuition), whichever side of the luxon wall you were on would seem to be the tardyon universe and it would always be the other side that was the tachyon universe. In that way, both sides would be tardyon to themselves; both sides tachyon to the other, and there would be perfect symmetry.

In the 1971 issue of the *McGraw-Hill Yearbook of Science and Technology,* Bilaniuk, in an article entitled "Space-Time," subjected the matter to careful mathematical analysis and found that there *was* just this symmetry between the two universes.

And if this is so, the speed limit remains. No matter how spaceships shift back and forth between universes, they are always tardyon and it is always the other universe that is going faster than light. Science-fiction writers must, after all, look elsewhere for their hyperspace.

I can't resist the temptation to include at least one essay that has never been published.

What happened is this: Some firm wanted to put out an information brochure on its activities. It produces chemical fertilizers, so it wanted to include, among other things, an imaginative look forward into the near future of agricultural technology.

They came to me for the purpose and I produced the following essay. They bought it and paid for it, but then the project folded for reasons that (I believe) had nothing to do with the essay.

So it has never been published—until now.

23

THE COMING DECADES IN AGRICULTURE

Four billion people on Earth today.

Seven billion, perhaps, in the year 2000. —Or will the extra mouths simply starve and will it be four billion still—if that?

We hope not. We hope that eventually the population boom will level off and begin a gradual decline to some optimum level. We hope that energy conservation and the development of new energy sources will keep technology moving forward.

And meanwhile, we hope to keep as many of the additional population alive as we can, while humanity solves its problem of numbers and energy.

But how?

First, can we extend the acreage that can be devoted to agriculture? Surely, this is not easily done. The naturally fertile land is already occupied, already sown; indeed, oversown, so that here and there the desert creeps in and cuts down the arable land.

But then, consider the desert. What makes a desert is,

commonly, lack of water. Earth, however does not lack water; indeed, it has a surplus of it. The trouble is that 98 per cent of the water is salty and most of the rest is polar ice. In either case, it is unavailable for agricultural use.

The ultimate agricultural water resource is rain, from which is born our streams, rivers, ponds, lakes, and groundwater. This freshwater supply of the world is used to irrigate arid land, but it can be used more efficiently; particularly the groundwater might be, for it underlays even desert regions if we dig far enough.

Furthermore, rain is not evenly distributed over the world, and this may be modified, too. Thanks to weather satellites, we already know more of global air and ocean circulation than might have been dreamed possible twenty years ago. In twenty years more, we may be developing the art of weather modification.

In addition, we may pass beyond the rain. There is the prospective use of icebergs as freshwater sources. There is ocean-water desalinization on a large scale.

But water need not be the only bottleneck that creates a desert. Plant life needs a variety of metals in trace quantities—metals that are not evenly distributed over the global surface. Areas unusually short in zinc, molybdenum, or copper, for instance, may have limited fertility even if water is supplied.

Again we can turn to our satellites to supply us with a global picture—for agriculture has become a global industry that cannot be dealt with efficiently in a local or regional way. It may be that in the next twenty years, fertilizers may be produced which match the needs of each unit of territory. Fertilizers may be turned out by recipe, so to speak, and will contain a careful mixture of trace metals required to produce maximum fertility in a particular area.

We will, in short, have the concept of reorganizing the soil to suit plant life, taking the various kinds of atoms from where they are oversupplied and placing them where they are undersupplied. Nor would this be a one-shot device by any means. Rain leaches minerals out of the soil; plants absorb them. The soil will have to be periodically readjusted.

Overall shortages of one element or another? Not likely. All elements find their ultimate home in the sea. There, they exist either in solution or in the form of metallic nodules scattered over the sea bottom and, in either case, can be withdrawn again.

The expansion of arable land and the adjustment of soil chemistry to the particular needs of our domesticated crops will, inevitably and of necessity, shrink the area of the Earth's surface available to the wilderness and to wildlife. This can't be helped as long as the human population continues to increase. But then, efforts to increase agricultural yield must be accompanied by efforts to halt the population increase or *any* measures to increase the food supply will, in the long run, fail.

Can the food supply be increased while expanding the arable land at the cost of the wilderness as little as possible?

For one thing, we might increase the number of plant species available for food.

Humanity mainly subsists on half a dozen or so varieties of grains, notably wheat, corn, and rice, but there are a number of tropical plants which could conceivably serve as a useful food source. In some places, it may be possible to grow them more abundantly than more conventional crops could be made to grow.

There could be aquaculture, too, in which seaweed is grown in shallow arms of the sea and there harvested.

With increasing knowledge of genetic-engineering techniques, new strains of plant species may be produced that not only grow faster, but will be richer in vitamin content and will contain proteins better balanced in the essential amino acids.

(Naturally, other strains, which may have advantages in weather resistance or disease immunity will be kept in being, for use in emergencies.)

Genetic engineering may also produce new strains of bacteria capable of fixing atmospheric nitrogen with greater efficiency, or capable of producing other desirable chemical changes in the soil. This would produce a continuing natural effect of greater fertility.

The fastest growing of all forms of life are the simple unicellular organisms. They can double their mass in hours or even minutes. Appropriate microorganisms might be cultured in mass to serve as food, or as food additives.

They would be grown in carefully balanced solutions designed to insure maximum rate of growth. We might imagine the preparation of a mineral mixture, dissolved in water, and added to the growing solution in a continuous drip at a known rate.

The whole affair might be automated with water, carbon dioxide, and minerals being fed in at one end and slabs of edible material emerging at the other. In between would be the busy cells, dividing and dividing and dividing again.

The cells may be adjusted by genetic-engineering techniques to produce food slabs with acceptable flavor. Failing that, the chemical industry will be further geared to the production of artificial flavors, aromas, and texturing agents.

This sort of food will, of course, be depressing to those who prefer the traditional foods of mankind, but if humanity allows its numbers to reach seven billion, traditional foods may simply not suffice.

Then, too, we can turn our attention to the never-ending war between *Homo sapiens* and other species for the same food supply. As human beings have multiplied the growth of certain selected plant species to serve as their food supply, competing plants have had to be uprooted and foraging animals have had to be eliminated. —Or at least, as much as possible.

This is not even a matter of interference with the ecological balance. It is the multiplication of these weeds and pests that is the interference (brought about, to be sure, by human activity).

The agricultural activity of humanity has created a new habitat made to order for certain kinds of plants and animals, which multiply enormously within that habitat as they never would have in a state of prehuman nature. Rats and mice would not multiply as they do if they had not adapted to a kind of parasitism on humanity. The agent that causes Dutch elm disease would not have flourished as it did if human beings had not planted elms row on row and made contagion simple.

The most sophisticated weapons in this battle against the competitors for food have been the chemicals used against the most persistent and dangerous of these competitors—the insects. Those chemicals have steadily increased in sophistication from crude mineral poisons deadly to life in general to organic substances that exert effects chiefly against insects.

Even the organic substances (DDT was the most famous) could have unwanted and even dangerous side effects. Moreover, they did not necessarily kill off all the insects. Those with natural immunity survived and gave birth to myriads with equal immunity so that resistance to any particular pesticide arises and spreads.

Those pesticides have been improved in both these respects; there are subtler methods that may come into use in the decades ahead. The complex life history of an insect involves the coming and going of certain hormones which mediate molting, chrysalis formation, and so on. These hormones are specific to insects and come in varieties which are each specific to a small group of species or even to one particular species—and affect no other living thing.

The use of the hormone itself or of its synthetic analogs may serve to upset the life cycle of a particular noxious insect without any side effects on any other species of living thing. Nor could an insect develop immunity to its own hormones.

Again, many insects mate successfully by having the male react to special chemical "pheromones" liberated by the receptive female. The male will respond automatically, though the substance has no effect on any other species. (If it did, its whole function of bringing about effective reproduction would be subverted.). Pheromones could therefore be used so as to mislead the males and prevent reproduction.

A victory over competing pests in agriculture could double the food supply of the world without putting one additional acre into agriculture and without growing one additional ear of grain.

Animal food is a luxury. It is always at least one step removed from the plant. If an animal is a herbivore, ten pounds of plant food will be consumed to bring one pound of animal food to the table. With the increase of the Earth's population there will be intensifying pressure to find ways of using the ten pounds of plant food directly for human consumption, rather than to use it to form one pound of animal food.

Nevertheless, the more that can be done to expand plant production, intensify its rate of growth, and rescue it from competing pests, the more room there will be left for animal production. In short, the future push toward vegetarianism, while growing steadily more powerful, may not overwhelm the human carnivore altogether.

Animal food production can be made more efficient, too. Cattle, for instance, eat grass and hay, which are high in cellulose, a substance human beings cannot digest.

But then, cattle cannot digest cellulose directly, either. They do so only through the presence of certain bacteria within their complicated stomachs, bacteria that *are* capable of digesting cellulose.

If, through genetic engineering, we produce strains of bacteria that can perform the cellulose-digestive process more quickly and efficiently, if we produce strains of bacteria that can break down other components of plant substances—we might envisage animals that live, literally, on sawdust and water to which small quantities of minerals and vitamins are added.

Then, too, the ocean can be turned from a food-hunting domain to a food-herding domain. Arthur C. Clarke, the science-fiction writer, once wrote a story about schools of fish being corraled by "cowboys" using dolphins in place of horses. Without going that far, there is no question but that ocean life might serve as a more efficient source of food if we could control the birth and upbringing of baby fish of desirable species to some extent.

And above and beyond all this is the prospect of farms in space; of space settlements with soils derived from the Moon and adjusted to ideal fertility; of farms, there, that have the controlled weather, moisture, and temperature that would be possible in an engineered cylinder, several miles long; of small agricultural worlds on which no noxious pests were introduced in the first place.

The bread basket of Earth may in the end be nowhere on its surface, but out there in space.

That, however, is for later on.

The responsibility for this article rests with Dan Button, the editor of Science Digest. *I had lunch with him soon after Jimmy Carter's election, and it occurred to him that since Jimmy was going to get advice from every direction, why not some from me? —Particularly if it were the kind of advice* Science Digest *could print in time for the inauguration.*

The thought amused me. Why not?

I was sure I could give the President some good advice, even though I was equally sure he wouldn't take it. The essay below was written, and did appear in the month of the inauguration. It has been reprinted in several places and has generated considerable mail, but I was right—there was no sign that anyone in the government would take such advice seriously.

Perhaps the catastrophe must approach more closely before we can be expected to begin to move in the right direction—after it is too late, of course.

24

AN OPEN LETTER TO THE PRESIDENT

Dear President Carter:

I am not one of those who thinks that a President is a miracle man who can, with a wave of his hand, change the world. There are limitations on the power of a President, ranging from the intractibility of the physical laws of the Universe to the stubbornness of public opinion.

Nevertheless, a President can push in a certain direction, acting where the Constitution and events permit and persuading where public opinion lags.

Let me begin by considering the matter of our energy supply. Americans use energy at a greater per capita rate than any other people on Earth and it is to this that we owe our high standard of living. This high standard has been built up because our territory has been rich in the prime energy yielders, coal and oil, and because our society has been of a nature that has made it possible for us to exploit these resources.

But oil, at least, is running out. American production peaked early in this decade and has since been inexorably decreasing. Even allowing for new supplies from Alaska and from the continental shelf, native oil will be virtually gone by 2000. Nor will imports help. Quite aside from the danger of growing ever-more dependent on foreign oil, world production will peak within ten years and will then begin inexorably to decrease.

We must, therefore, as you have said on numerous occasions during your election campaign, develop a rational energy plan for the nation.

We can conserve our present energy supply, and absolutely must do so, but, with the best will in the world, we can only conserve so much, so conservation will only stave off the evil day a couple of decades.

We may turn to greater use of coal and to the use of oil shale, but both will have serious environmental consequences. We may develop more sophisticated uses of such old standbys as wind power and waterpower and such new developments as tidal power and wave power. This will give us additional time—but again only a limited time, and by themselves they are not enough.

We can grow plants for alcohol as fuel, but these will eventually compete for space with plants grown for food, and we will face a struggle of competing pressures of the two kinds of starvation.

We may rely more heavily on nuclear-fission power, but you have training in this field and know the risks and dangers involved. We may pass on to nuclear-fusion power—far more plentiful than fission and, possibly, far less dangerous—but controlled fusion has not yet been developed, and we cannot yet be certain that it will be.

We may pass on to geothermal energy and to the direct use of solar energy. Both have promise, but both will require great capital investments.

It is quite possible that by a mixture of all of these, the United States, and mankind generally, can pull through, but there is one more way of getting energy that is, in my opinion, better and more valuable than any of those I have mentioned:

Direct solar energy as received on the surface of the Earth is blocked to some extent by the atmosphere on even a clear day. Where there is special atmospheric interference in the form of clouds, mist, and fog, there is major blocking. And, of course, it is blocked altogether at night.

Then, too, because the energy of sunlight is dilute and because conversion devices are inefficient, it must be collected over a large area if it is to be useful as a major energy source. Several thousand square miles of southwestern desert area (where sunlight is most reliable) would have to be coated with solar batteries.

Why not, then, make use of space? Why not have several solar power stations placed in synchronous orbit about the Earth, some 22,000 miles above the surface, each one moving so as to stay, more or less, over a single spot on Earth's equator?

A solar power station can receive the full range of the Sun's energy, unblocked by atmosphere and atmospheric phenomena. It will be in Earth's shadow only briefly each night at the time of the equinoxes; not at all at other times, and when one is in shadow others will be in sunshine. It can convert solar energy to a beam of microwaves which can be picked up and used with much greater efficiency than sunlight itself can be, so that the collecting areas on Earth's surface will be much smaller and more easily maintained.

Finally, energy stations in space will give us something much more important even than energy:

The world, at present, is divided into well over a hundred competing nations, each one of which considers its own particular needs and desires to be paramount. At least two of these nations have the capacity to destroy civilization in a matter of hours if they should choose to go to war. Even if there is no war at all, the resources and energy spent on maintaining competing war machines is something the world can no longer afford.

Aside from the material danger, antagonism and rivalry among nations is no longer thinkable because it consumes efforts of reason, outpourings of emotions, intensities of drive and ambition that are uselessly wasted when they are devoted to such ends. There is only one war that the human species can now afford to fight and that is the war against extinction. It is the effort for human survival that should claim all our reason, emotion, and drive, because all we can supply will be needed if survival is to be achieved. Any lesser struggle is trivial and will, in any case, be doomed as soon as the greater struggle is lost.

Nor is this war for survival a matter only for our own nation to be concerned with. All the great life-and-death problems that now affect the United States affect the world generally. The problems of overpopulation and soil deterioration, the problems of resource depletion and of pollution; the problems of alienation and terrorism—these affect all the globe.

Nor can these problems be satisfactorily solved by any nation within its own boundaries, however large, wealthy, or populous that nation might be. The Earth itself is now too small, and the nations too interconnected, economically and ecologically, for any solution to make sense on a purely national basis. There will have to be international cooperation, closer than any that yet exists, if these problems are to be solved.

As the years pass, it may well be that the intensification of these problems will force a form of such cooperation, but it is likely that mutual dislikes, rivalries, and antagonisms may well reduce the efficiency with which solutions can be arrived at and implemented, and do so to the point of absolute failure.

Somehow, it will have to become clear to the nations of the world—and, even more important, to the *people* of the world—that *real* cooperation among the nations will bring about enormous benefits that are greatly to be desired, and that these will be lost otherwise.

And there we come back to the notion of power stations in space.

All forms of energy obtained here on Earth are geography-bound. Just as there are oil-rich regions and oil-poor regions, so some regions are richer than others in tidal energy, in water power, in geothermal energy, in uranium supplies, in access to the ocean. And while that is so, the temptations to rivalry may be impossible to defeat.

Space, however, is equidistant from all surface regions. Energy drawn from the Sun in space would belong to all people in a fashion that no earthbound energy source possibly could. The power stations in space would, in a sense, breed cooperation.

Furthermore, the task of building such space stations is so enormous that it could well be made an international project.

Not only would this reduce the cost to the United States alone, but it would give the peoples of the world a task (one of immediate profit to themselves) to which they could all contribute. It would capture their imagination, lift their pride, raise their hopes. The construction and maintenance of such stations would be so global and human a project that it would dwarf and make clearly disreputable any petty ambitions on a smaller scale. What's more, once the stations are built, it would be clear that if humanity falls into its earlier rivalries, those stations might well fail through lack of proper maintenance. That would provide an overwhelming incentive for cooperation to continue.

The techniques developed for the building of such stations could be used to build other structures in space. There could be space observatories for the study of astronomy and other sciences. There could be space laboratories where dangerous experiments in nuclear physics and in genetic engineering could be conducted with little risk to Earth itself. There could be factories that would take advantage of the peculiar properties of space, such as high vacuum, high and low temperatures, hard radiation, to carry on industrial procedures and to produce devices difficult or impossible to carry on or produce in Earth's surface environment.

As more and more is done in space, it will be increasingly useful to get rid of a wasteful commuting problem by placing more and more people in space more or less permanently. We would, in short, build space settlements.

There, material for building these observatories, laboratories, factories, and settlements could be obtained almost entirely from the Moon. For the foreseeable future, at any rate, the Moon would be an inexhaustible source of material of all kinds (but for hydrogen, carbon, and nitrogen). What's more, the Moon is without a native ecology we might hesitate to disturb.

Three great ultimate consequences may result from our space activities, aside from obtaining energy and gaining world cooperation:

1. Eventually, there might be enough space settlements to permit a new growth of human numbers, after that growth has come to a halt (as it must in the near future) on Earth's surface.

2. Eventually, more and more of Earth's industry will be lifted into space, where the problem of resource depletion

(thanks to the Moon) and of pollution (thanks to the great volume of space) will be far less important than on Earth. With industry on Earth's surface shrinking, that surface might return more nearly to a wilderness/park/farm condition that most people would find more desirable than the present condition. What's more, we would be restoring the beauty of the Earth without losing the material advantages to ourselves of industry and high technology.

3. Eventually, the people of the space settlements would serve as the advance guard of humanity in the exploration and settlement of the farther reaches of space. People, living enclosed in small worlds, and accustomed to space travel, would have far fewer and far less intense psychological problems in making long space flights than would people who have been brought up under the space-alien conditions of the Earth's surface.

The importance of this can be seen if we remember that the advance of humanity till now has come about, after all, through the steady increase of its range over the face of the Earth, the steady increase of the sophistication of its methods for transportation and communication, the steady increase of its ability to make use of the laws of nature for its own purpose. But the Earth is filled now; overfilled indeed; and we can do nothing more if we must remain confined in the strait jacket of the planetary surface. Indeed, we can only decline and perish.

We must therefore expand again—toward the unbounded horizons of space.

The totality of this vision cannot be brought about in four years, President Carter; or even in eight years, assuming you are reelected in 1980; but a start can be made. The vision can be presented as a goal to the people of the United States—no, to the people of the world—by yourself far more effectively than by me.

You can become space-oriented then, President Carter; you can begin the planning, on an international scale, of the steps by which we may safely take what I believe to be the only road that will lead to the salvation of civilization.

Few of my essays have had a more curious origin than this one.

Bernard Schwartz of New York University was organizing a colloquium on the law, in order to celebrate the Bicentennial, and wanted a contribution from me. My natural protestation that I knew nothing of the law went unheeded, and the pressure continued, along with heaping quantities of flattery (to which I am very susceptible).

Not only was I compelled to use my ingenuity in thinking up a way of complying with the request—that I look at the future of law from the perspective of space—but I even had to come to New York University and deliver the essay in the form of a speech.

Well, I can't expect life to be nothing but roses.

25

SPACE AND THE LAW

When asked to write this paper, I stated, quite frankly, that I knew nothing about the law and that I asked nothing better of life than to be forced into as little contact with it as possible. This objection was brushed aside, and I seem to have gained the impression from what was said that I would not suffer, in this respect, by comparison with many of the other learned gentlemen participating in the conference.

Despite this possible community of ignorance (if, indeed, one exists), I am not so naive as to let myself be lured into a discussion of the fine points of the law, as I imagine them, or into a comparison of sea law and space law, for instance. I must leave this to wiser heads, or, at any rate, more specialized ones. I will confine myself instead to my own specialty—which is the constructions of visions of the future.

Such visions may prove to be meaningless, but those who

organized the conference were warned of this by me and yet insisted, nevertheless, that I continue. My conscience is, therefore, as clear as a mountain spring.

Space—the vast realm beyond Earth's atmosphere—cannot have a significant effect upon human society and upon the local rules of conduct on which it is based, as long as human beings have not invaded it in a significant manner.

What has happened, so far, has been the temporary orbiting of the Earth, by ships carrying one to three human beings, at varying distances above the surface; plus six separate, but brief, visits to the Moons. These are *tours de force* and, as yet, nothing more. They have not yet given rise to the overwhelming sensation that new laws are needed or that old laws, based on the prespace dispensation, are outmoded.

Things will change when human beings move out into space more or less permanently; when the Moon becomes a source of material over which human beings can quarrel; when the Moon or artificial structures in open space become suitable as the site of industries, laboratories, or homes (or all three) under conditions so alien from those prevailing on Earth as to make Earthly law perhaps no more than a dim and, if anything, misleading precedent.

It is very unlikely that this can happen very soon, however. Unfortunately, there is a rising crisis on Earth generally that makes it less probable from year to year that resources, capital, and effort will be put into building habitats in the sky, as long as the habitat on Earth continues to deteriorate. Our space effort will, in the short run, dwindle, and the question of the interconnection of space and the law will fade in consequence.

The long run is another matter, assuming that as far as civilization is concerned there *is* a long run. The rising tide of population and the dwindling ebb of resources will face humanity, alas, with the specter of starvation—ourselves for food, our technology for energy and basic materials, our environment for viability.

Civilization may survive the crisis (and how awful the word "may" is in this connection) but, if so, what will emerge in the twenty-first century will have to be quite different from anything we have seen before in history. The twenty-first century will have to be a low-birthrate society if civilization is to make it, for population must be stablized and, very likely, must be lowered in, we hope, a slow, steady, and humane fashion.

The effort to correct the cancerous growth of population and the open-ended ravishment of Earth's resources, together with all the ills this has brought about, will clearly have to be global in its nature. The problems created by the very success of human technology, human medicine, and, therefore, human proliferation are impossible to confine to one section or another of the globe. All the life-and-death problems of the world—population, pollution, scarcity, and war—are, indeed, of the *world*, and there are no havens of refuge or avenues of escape.

Nor can any section of the world solve its own problems at the expense of its neighbors. No section can flourish by looting another. However much against our will it may be, it will turn out that human beings, embedded in a globally interconnected economy, can flourish individually only if all human beings flourish generally. The kick you aim at your neighbor will hit your own rear as well.

In short, then, if civilization does survive intact into the next century, it will be because there will be a kind of global organization, a kind of international cooperation, that will consider global problems, weigh possible solutions, decide on the most useful, enact them and enforce them. I don't know what to call this except "world government."

The concept of world government is not popular anywhere on Earth except in the case of occasional ivory-tower idealists like myself. Unfortunately, most people are quite convinced that they are superior to their neighbors (especially if their neighbors commit the unconscionable crime of looking, speaking, or behaving differently from themselves) and naturally resent placing those neighbors in a position to help determine the rules of society.

Even when vile necessity forces cooperation (as when wartime allies, with sidelong glances and under-their-breath mutterings, are reluctantly forced to march in unison against a formidable enemy), the slightest easing of the situation will produce instant enmity among the erstwhile friends.

If the survival of civilization is to be more than a perpetual drunken stagger at the brink of the bottomless pit, the global community will have to be firmer and more steadfast than that. The great legal problem of the twenty-first century will, then, have to be the establishment of a strong world government, capable of withstanding the centrifugal shocks of separatism, and yet one that will not drown out regional self-rule in matters that are not of global interest.

But how can that be accomplished?

One way, surely, could be to give the people of the globe some project that is large enough and glamorous enough to engage their hearts and minds; something to which all can contribute in one way or another and in which they can feel themselves to be global citizens, members of *Homo sapiens*, rather than members of this or that region inhabited by this or that subspecies.

In the past, we could sometimes rely on a war crisis, for then it might happen that all questions of party are submerged in an orgy of self-immolation for the motherland or the fatherland or the grand old flag—or whatever combination of syllables can most easily tap reflex emotions.

War and civilization have become incompatible, however, and we must have something nonmilitary and constructive for mankind to rally round.

It seems to me that the obvious project, large and glamorous, clearly global in scope, is that of colonizing space—of building cities on the Moon or of building habitable doughnuts of glass and metal in open space.

For humanity to expand into space would entail enormous profits of all sorts.

It would enlarge knowledge, with incalculable and largely unforeseeable consequences, which we can hope that human wisdom (not always a markedly visible commodity) will guide into good and useful paths.

It would make possible industry and research of a kind that would profit in an environment that would include (depending on time and place) high vacuum, deep cold, or intense radiation, thus multiplying human ability to cooperate with the laws of nature.

It would lead to the construction of power stations in synchronous orbit that would collect solar energy for human use with far greater efficiency than would be possible on Earth's surface, and thus solving our present problems of energy shortage and of pollution—or, at any rate, ameliorating them.

All these material benefits, and others, would become so obvious so soon that people will stare in disbelief or laugh in derision at the twentieth-century accusations that space exploration is but a vast boondoggle, and those public men who will have committed quotes to that effect will achieve an unenviable immortality as comic examples of shortsightedness.

Yet it is not with material benefits I wish to concern myself here, but rather with something more. After all, nothing material space can bring us will be as great in value as the immaterial benefit of increasing humanity's appreciation of itself as a global set and as a single species; and of strengthening world government and making it workable—and therefore making civilization workable.

But can colonization possibly strengthen the concept and practice of world government and can it make possible the existence and development of global law?

If we eliminate simple assertion in one direction or another as a fruitful way of arriving at a clear view of the future, we are compelled to argue from analogy and to look back in history for incidents that might illuminate the future by way of the past.

When we think of colonization today, we are apt to think of a technologically superior society establishing itself on land occupied by "natives." A thin outpost-layer of the colonizing society then oppresses and exploits the "natives," growing fat at their expense. This type of colonization has run its course and is viewed, now, with opprobrium.

This sort of colonization cannot, however, be viewed as analogous to the colonization of space where (at least in the immediate neighborhood of Earth, and perhaps in the entire Solar system) there is no intelligent life, or life of any kind, to displace, mistreat, and exploit.

Must we then look to the past for occasions when some society occupied an empty land (at least as far as intelligent life is concerned) and established extensions of itself?

Such a search is not likely to be fruitful. The last occasions when this happened on a continental scale were when the forebears of the Indians passed from Siberia into Alaska and spread over the American continents, and when the forebears of the Aborigines traveled through the Indonesian islands from Asia into Australia. Much later, the same happened on an oceanic scale when, in the first millennium A.D., the Polynesians scoured the Pacific Ocean and settled the scraps of land they encountered.

These colonizers, however, possessed "prehistoric" societies and had not yet developed cities and writing, those hallmarks of what we call "civilization." We don't know enough detail about such prehistoric colonizing ventures to shed the necessary light on what might happen in the case of space colonization by societies that have reached such a pitch of civilization as to be virtually on the edge of suicide.

But there have been no *civilized* colonizing ventures of consequence that have ventured into empty lands. By the time civilization had arisen in the Middle East, there were no empty lands within reasonable reach of that civilization.

The next best thing is to find colonizing ventures which created their own empty land by pushing "natives" back, or exterminating them, or both. However morally reprehensible this is, the consequence is that the colonizing society produces a colony like itself without a significant residue of "natives" to exploit and, in this respect, can be considered a possibly useful analogy to space colonization and its consequences.

Let us consider the Greeks of the first millennium B.C. It is from the Greeks that the modern Western world traces its philosophy, art, literature, science, and even important strands of its religion. Generally, we of the West feel sympathetic to the Greeks and identify with them in their struggles against the non-Greek "barbarians," notably in their war against the Persian Empire.

Yet the ancient Greeks were not, at any time in their history, a single realm under a central government. They were a congerie of city-states, whose normal situation was that of mutual suspicion and hostility, and who would turn on any one of their number who showed signs of growing too powerful.

They could never, under any circumstances, truly unite against an external enemy. Their closest approach to this was in the war against Persia between 500 and 450 B.C., and even then, substantial portions of the Greek world remained neutral—and in some cases even sided with the enemy.

This was true despite the fact that the Greeks recognized themselves to share a common heritage, a common language, a common literature, a common religion. It was true despite the fact that the Greeks recognized themselves to be a unity at least to the extent that they lumped all non-Greeks together as "barbarians."

After the successful fight against Persia, the Greeks could not maintain even the limited union that marked that war and, through division, fell easy prey first to Macedonia and finally to Rome.

Was there no great accomplishment behind which a pan-Greek spirit might have flourished?

Yes, there was, and it was even a great burst of colonizing energy that in some ways could offer an analogy to the projected twenty-first-century colonization of space.

The Greeks, between 750 B.C. and 550 B.C., expanded greatly, planting colonies all along the Mediterranean coast from the farthest eastern reach of the Black Sea to the coastline of Spain in the far west, pushing back the non-Greeks already on the spot and constructing extensions of their own culture.

Yet the period of colonization did not create the necessary spirit of union. For one thing, each colony was an independent city-state which promptly took up the task of fighting its city-state neighbors to the point where all could barely hold their own against the competing imperialisms of the Phoenicians, Carthaginians, Etruscans and, eventually, could not hold their own at all against the Romans.

Then, too, each colony was the product of the colonizing venture of a single city-state, so that Miletus was the colony of Athens; Syracuse of Corinth; Byzantium of Megara; Taras of Sparta, and so on. Many of the colonies then, in turn, founded colonies of their own.

Any political or emotional ties that the colonies might form were therefore never with the Greek world as a whole, but at best with the mother city. The result was that when cities warred, each sometimes looked to a mother or daughter city for help, and disunion was exacerbated.

A curious parallel to the Greek experience was that of the great colonizing ventures of Western Europe between 1400 and 1800 A.D. As was the case with the Greeks, Western Europe had never been unified at any time after the fall of the west Roman Empire in the fifth century, except for a partial and abortive union under Charlemagne about 800 A.D. This disunity, as in the earlier case, persisted despite a common heritage, a common language of learning, a common literature, and a common religion.

Nor did union develop through the great adventure of colonization.

Some of the colonization, it is true, took place in Africa and Asia and was of the sort that established the domination of a European minority over a non-European majority. If we ignore this as an aberrant form, there remains the colonization of the Americas and of Australia where the native inhabitants were pushed back or destroyed and in which extensions of the colonizing culture were established as had been the case earlier, in the Greek colonizing period.

In the European case, unlike that of the Greeks, the colonies were not independent from the start. Instead, each European nation kept its colonies tightly bound to itself and exploited them economically. (Eventually, these colonies rebelled and broke away, of course.)

As in the Greek case, however, the colonies were established by single political units of the overall culture. No colony felt any allegiance to the European world in general, but only to individual nations. As a result, the colonies did not encourage union but rather exacerbated disunion, and colonial rivalries among the great colonizing powers became a new occasion for the endless wars that racked Europe as once they had racked Greece.

But now let us take a third case. In the period from 1600 to 1750, England (later Great Britain) had planted a series of colonies along the east-central coast of North America (pushing back the Indians and eventually destroying them in casual genocide).

These colonies had varying degrees of self-government, but whatever their ties to the mother country, they were certainly independent of each other. There was no way in which the people of Massachusetts could help make the laws that governed Virginia, or vice versa.

In the Revolutionary War of 1775–83, under the stress of the struggle for independence, the colonies joined in a weak and shaky alliance no stronger than the union of Greek city-states against Persia. Nor was it a unanimous alliance, any more than the Greek one had been, for the colonists in Nova Scotia, Newfoundland, upper and lower Canada, and the West Indian islands did not join the rebellion. In fact, they wholeheartedly supported Great Britain. What is more, even within the thirteen colonies which were nominally in rebellion, at least as many colonists fought for Great Britain as for independence.

The war ended at last with American independence established but that did not make the alliance of thirteen self-governing and essentially independent "states" any stronger. Few political observers in Europe thought that the "United States" would remain united for long.

There are a number of reasons why the Union persisted, but one of them rests with a crucial self-abnegating decision on the part of the infant states—a decision whose wisdom and critical effect has been underestimated by posterity.

The new states were primarily settled east of the Allegheny Mountains, but the territory of the new nation stretched westward to the Mississippi River. The royal charters that had originally established the colonies-turned-states were vague as to boundaries and generous in granting everything westward to the setting sun. The result was that nine of the states had claims, conflicting and overlapping, to the western lands. The land north of the Ohio River, for instance (the Northwest Territory), was claimed entirely by Virginia and, in part, by Pennsylvania, Connecticut, Massachusetts, and New York.

Had such conflicting claims persisted, there would have been an endless cause of wrangling among the states, and the earlier Greek and European experiences would have been relived. Had the claims been straightened out and the western lands distributed among the states according to some compromise, conflicting imperialisms would nevertheless have been set up, and each further extension of national territory would have been the occasion for endless rivalries. The nation in the end would have become a congerie of battling subnations.

What really happened was that one of the states *without* western claims, Maryland, refused to join the Union even under the weak terms of alliance that had existed during the Revolutionary War until all western claims were abandoned and the western lands turned over to the weak and almost powerless Congress.

One by one, to appease Maryland, the various states abandoned their western claims.

Not only was a cause of rivalry removed, but the central legislative body, the Congress, had in this way gained something. Congress was so weak in the years following the Revolutionary War that it was almost a negligible quantity, as the United Nations is today. Like the United Nations, the early Congress would not levy taxes but had to wait, hat in hand, for contributions from the states that made up the Union—contributions that came reluctantly, sparingly, or not at all.

But Congress now had the western lands! What was it to do with them?

On July 13, 1787, with only eighteen members of Congress present (so moribund was the body), the "Northwest Ordinance" was passed. By that Ordinance, it was decided that when population reached a certain level, new states would be formed out of the Northwest Territory and these new states would be equal to the old ones in all political and social respects. No state would be the superior of another because it was older or because it had been one of the original thirteen. Had this point not been made clear, the Union might have become a mixture of dominating senior states and dominated junior ones, and the stage would have been set for new rebellions.

The new states, like the Greek colonies, had equal status with the mother states, but unlike the Greek colonies maintained strong political ties with the colonizing power. These strong, political ties were not, as in the case of the European colonies, to a single segment of the colonizing culture, but to the culture as a whole. The new states became part of a central government which, in 1787, was made infinitely stronger by the working out of a federal constitution which was quickly adopted by the various states. (This adoption meant that the states voluntarily surrendered key portions of their sovereignty—again an enormously wise example of self-abnegation.)

There was another reason why American colonization, in the form of new states, strengthened the Union—unlike the case of the earlier Greek and European examples.

The new American states were not officially founded by particular old American states. Most of the immigrants would naturally come from nearby states, but anyone could enter from any state—and did. The result was that, on the whole, blocs of states were not formed. There was not, from the start, a Virginia bloc and a New York bloc and so on.

The discords and dangers to which that would have given rise are clearly shown through the one case where blocs *did* form, but in a manner that did not arise out of the way in which the new states were formed.

Unfortunately, the states while they had still been colonies, had imported black slaves from Africa and *created* a master-slave society where none need have existed. The unwisdom of this decision went far to neutralizing all the wisdom that had gone into the establishment of a working Union.

As time went on, some of the states outlawed slavery and a growing hostility arose between them and the other states that permitted it. Unfortunately, the two groups were not scattered, but formed two compact blocs.

The consequences showed what might have happened if the United States had been broken up into half a dozen spheres of influences. In 1859, when Kansas was approaching entry as a state, possibly free and possibly slave, both sides attempted to influence the vote by subsidizing immigrants, arming them, and encouraging force. The result was a civil war in Kansas.

Within two years, there was a Civil War in the nation as a whole.

The United States survived, at great cost, partly because there were only two blocs. One could be defeated, the other be victorious. Had there been half a dozen blocs, the shifting alliances that would have resulted as each bloc perceived danger to itself every time a competing bloc seemed to grow too strong would have made a clear-cut conclusion impossible and the United States would have disintegrated.

What was more amazing than the victory of the Union in 1865, however, was the fact that it was followed by reconciliation. The defeated slave states, prostrated and humiliated, did not easily forget or forgive and haven't entirely forgotten and forgiven to this day, but the spirit of revenge was contained. The embittered memory of defeat did not result in further revolts or in a guerrilla movement or a long-lived independence drive or appeals to foreign powers. Instead, reconciliation while slow, was real, and the United States has remained strong and united.

How did this come about? Partly the responsibility must surely rest with the historic accident that, in the decades immediately after the Civil War, what remained of the West was developed and a dozen new states were formed—states which were settled by men from the victorious North and from the defeated South on an equal basis; states which, therefore, owed allegiance to neither of the two sides but only to the nation as a whole. In the great task of the colonization of the West, the sectional wounds were healed.

* * *

If we agree, then, that the Greek civilization and the West European civilization each harmed itself enormously through an inability to unite, while the American civilization developed unprecedented power-plus-liberty over a vaster area than had ever before been the subject of such an experiment—and if we further agree that the Greek and European systems of colonization *did not* further union at home, while the American system of colonization *did* and was perhaps the key to American strength—then what can we say about the forthcoming colonization of space in the twenty-first century?

1. The colonizing culture should be united to begin with, as Greece and Europe were not and as the United States was, even though the union might be excessively weak. This we may fairly hope for, since twentieth-century civilization is not likely to survive without some form of global cooperation and since a model for it already exists in the form of the United Nations (though one could scarcely imagine a more feeble and ineffective global government).

2. The space colonies must not be utterly independent as soon as they are capable of such independence, as in the Greek case; nor must they be subjected to such humiliating dependence as to be forced into rebellion and consequent utter independence, as in the European case. Either way, the results will not be conducive to global government. Instead, the space colonies must be tied to earth but under terms which give them the full rights and privileges of the Earthly units themselves, as in the American example. This we may hope for, too, since the space colonies will not be sufficiently self-supporting for a period of time and can scarcely hope for complete independence very soon, and since colonial oppression had gone out of style and, we may hope, will remain out of style.

3. The space colonies must not feel bound, either politically (as in the European case) or emotionally (as in the Greek case) to one single colonizing unit of the greater whole since that would encourage rivalry and disunion; but must feel bound to the central government only (as in the American case).

This third requirement is the most crucial since it seems the least likely. To make it possible, the global government must be in charge of the colonizing ventures, and the colonization must take place under global auspices. Each colony must be open for colonization, without restriction, to people from any part of the Earth, and, indeed, a well-mixed population should be positively encouraged on every colony.

We should avoid, like the plague, the formation of colonies populated entirely by Americans, or by Russians, or by Uruguayans—or by any group that would then feel some emotional or traditional attachment to one section of the Earth more than another.

A well-mixed colony would, instead, be a microcosm of Earth and would be divorced from the local rivalries of the civilizing power. Naturally, we cannot expect perfection. On any of the colonies there will be occasionally fashionable nostalgia that will make itself felt in the boosting of various ethnic heritages. We are experiencing this today in the United States, in fact, and it is a far cry from this to the deadly rivalries that result when opposing groups are armed and are ready to convert hatred into force.

And then, as the number of colonies increases, each with problems having nothing to do with terrestrial localisms, those localisms will seem to grow progressively more picayune and meaningless, and the global government will grow ever stronger and more meaningful.

My conclusion, then, is that if the colonization of space is carried through as wisely and as farsightedly as the colonization of the American West, it will be a vast project that will unite humanity both in the performance and in the consequence, and may be the route by which we can establish a functional global law for the first time in history and, in consequence, make human civilization permanent.

So to those who cry out that space exploration is too expensive, I can only ask: How much is survival worth?

It seems appropriate to end with this essay, which represents as far-flung and as wide-ranging a look into the future of man, of life, of everything, as I have ever managed.

26

A CHOICE OF CATASTROPHES

If one decides the world has a beginning, then surely it must have an ending, too. Generally, if it is thought the world began not very long ago, it is natural to suppose it should come to an end not very long from now.

The somber mythology of the Norsemen saw an end to the world that was not to be long delayed, for instance. There would come Ragnarok, the Twilight of the Gods, when the gods and heroes would march out to meet their deadly enemies, the giants and monsters, in the last climactic battle that would destroy the world.

Similarly, the Bible, which tells of the beginning of the heaven and the earth in the first book of the Old Testament, speaks of the end, too, in Revelation, the last book of the New Testament. It tells of a climactic battle at Armageddon and a final Day of Judgment.

The Bible does not give the actual time of the Day of Judgment but the early Christians appear to have expected it to come soon since the Savior had appeared and completed His mission. The end did not come, but in each generation there were those who proclaimed it imminent.

The year 1000, when it came, brought panic to some parts of Christendom, for Christians read into the Revelation reference to the thousand years that preceded the end, the thousand years that had just passed. When the end did not come, other calculations were made—over and over.

In the 1830s, a New York farmer, William Miller, calculated

the end would come in 1843, and many people sold all they had, donned white robes, and waited on hilltops. Nothing happened, but the movement gave rise to the Adventists, who still wait.

In 1879, Jehovah's Witnesses came into being as an offshoot of the Adventists and they waited for an imminent end. They still wait. They are still sure it is imminent.

Others have expected more secular ends. Comets have always been feared as omens of disaster and destruction, and as late as 1910, when Halley's Comet made its most recent appearance, uncounted numbers feared that Earth would be destroyed in its passing. Just a few years ago, there were foolish irrationalists who predicted that the passing of the small planetoid, Icarus, would cause California to fall into the Pacific Ocean.

Nothing ever happens, but those who wait for the end to come are never discouraged but are always ready with a new prediction.

What then can we say about catastrophe? With the view of the world as science has given it to us these last three centuries, can we laugh and say that the world will not, and cannot, come to an end?

No, for science treats of the beginning of planets and suns and of the whole Universe and, therefore, must treat of the end as well. And, indeed, there are ways in which the whole Universe might be considered as ending and that would mean the end of its component parts, of the Sun, of the Earth, of life upon the Earth (assuming that these have not ended already long before the Universe did).

For instance, we know that the Universe is expanding and that it may do so forever. As it expands, the individual stars of which it is composed consume their fuel and finally can radiate no more. The birth of new stars is no longer possible when all the hydrogen (the basic fuel in the Universe) is consumed. The Universe will then have run down, and if the end for us has not come before, it will surely come then. Still, the general rundown of the Universe will not happen for many trillions of years and it therefore need not concern us. There are other catastrophes, less all-embracing, but ones that will serve to end us, that will come before then.

Accompanying the expansion of the Universe is the possibility of a change in its fundamental laws. Some scientists speculate, for instance, that the force of gravity is slowly weakening as the Universe expands. This has the potential of catastrophe for us except that, even if gravity does weaken, or if other such changes take place (and this has not yet been demonstrated), it would take a billion or more years for the effect to become noticeable, let alone catastrophic.

But then, the Universe need not expand forever. Under certain conditions (and astronomers are not certain whether those conditions are actually met or not) the expansion would be slowing continually and will eventually come to a halt. The Universe would then very gradually begin to contract again, contracting faster and faster—and winding up as it does so.

This, however, does not give us a new lease on life, for a contracting Universe pushes radiation ahead of itself into a more and more energetic form that would be fatal to all life. However, even if the Universe does enter into a contraction cycle, that will not happen for perhaps 25 billion years and there again we are in no immediate danger.

It may be, too, that superimposed on the general expansion of the Universe are local contractions brought about by violent events in the history of giant stars and giant clusters of stars. These contractions may force matter together so densely as to form "black holes" from which nothing can emerge. It may be that black holes already exist and that they continually grow and encroach on matter until everything is gone.

When will a black hole engulf us? That depends on where one is located with respect to us. Are we on a collision orbit with one? Astronomers have detected a few objects they suspect might be large Sun-sized black holes, but they are so far away that again possible collisions must be billions of years away at the very least.

Some astronomers suspect that black holes can come in all sizes, down to very tiny ones no larger than atoms. There is even a suggestion that the 1908 incident in Siberia in which a whole forest was leveled, without any signs of a meteorite strike, was the result of the passage of such a mini-black hole into Siberia, through the body of the Earth, and out into the Atlantic Ocean.

Black holes are difficult to detect. Might there be one small enough and far enough not to detect at the moment, yet large enough and near enough to destroy our Earth on collision, say, a thousand years from now?

Here, however, we are talking of impalpables. Astronomers may speculate that small black holes exist, but they have detected none and they may not exist. And if they do exist, there is no way of knowing whether any are on the outskirts of our Solar system or not—and there is no reason, whatever, to think any are. We can only dismiss the possibility with a fatalistic shrug and a hope that, as we learn more, we can better study the nature of surrounding space and see what dangers lurk there.

If we assume that events outside the Solar system involve only catastrophes that can't affect us for many billions of years, or whose coming is utterly unpredictable, then we are left to wonder whether anything can bring us to a general end that involves our Solar system only.

To begin with, there's our Sun. If we are content to admit the Universe won't last forever, surely our Sun won't. In fact, our Sun should last a far shorter period of time than the Universe does. The Sun has now been shining at the expense of its hydrogen fuel for some 5 billion years or so. Eventually, that fuel will run sufficiently low to bring about changes in the solar interior that will cause the Sun to swell up into a red giant. When that time comes, the Earth will heat up to the point where life will not be possible upon it.

This, however, is not expected to happen for some 8 billion years and by that time humankind or its descendants (if they have not been killed off by some other prior catastrophe) may have moved on to other, younger stars.

Even if the Sun remains in its present form, might there not be minor variations, insignificant to the Sun itself, but deadly to the Earth? Might there be changes in the Sun's spot cycle or in its interior that will cause it to warm slightly or cool slightly—*slightly,* but enough to boil our oceans, or freeze them, and in either case end life on our planet?

This is not likely. The geological record (and the fossil record, too) would indicate that the Sun has been fairly stable for billions of years and it ought, therefore, to be stable for billions more.

What about the rest of the Solar system? Is any part of it going to crash into us?

Velikovsky and his followers believe that in the recent past, only 3,500 years ago, Venus, Earth, and Mars kept undergoing near collisions. Rational astronomers find it impossible to take this seriously. There is every indication that the Solar system is dynamically stable, that the major planets have kept to their orbits for indefinite millions of years in the past and will continue to do so for indefinite millions of years in the future.

But the Solar system is filled with debris: with minor planets (asteroids) of all sizes, from a few that are hundreds of miles across to many thousands that are only a few miles or even a few hundred yards across. There are uncounted particles ranging in size from a few feet across down to microscopic bits of dust. Some of these minor bodies are close at hand. There are asteroids a mile or two across that have orbits that can occasionally place those asteroids within a few million miles of Earth. And there must be bodies that are smaller still that can come closer—that can even collide with us.

Tiny micrometeoroids are indeed constantly colliding with Earth by the millions and burning up in our atmosphere (the larger ones visible as "shooting stars"). Particularly large meteoroids, from several inches to several feet across, can survive to strike the earth as "meteorites" and these can do some damage if they score a direct hit on human beings or their works.

The larger the meteoroids the more damage they do, and in primordial times, there was a larger number of quite large bodies. The craters on the Moon, on Mars, on Mercury, on the satellites of Mars and Jupiter were caused by collisions with sizable bodies. As a result, almost all of them have been swept up and interplanetary space is almost clean.

Almost, but not quite. A few sizable bodies remain. One left a crater in Arizona over half a mile across and it was formed, perhaps, ten thousand years ago. There are signs of other craters, some even larger, formed longer ago. Nowadays, the collision of a meteoroid capable of forming such a crater might wipe out a city if it happened to follow a course of unfortunate accuracy. A similar meteoroid striking the ocean might cause a splash that would devastate the coastlines of the world with towering waves of water.

What are the chances that there will be a devastating meteoroid strike in the near future? How can we say? On the one hand, there are fewer such bodies in space now than ever before, since every one that strikes is one less. On the other hand, there is a much greater chance of a meteoroid doing damage on our teeming, artifact-covered Earth now than there was of affecting our much emptier Earth of only a few centuries ago.

All we can say is that a strike *may* come tomorrow and *may* have a most unfortunate aim, but it is more likely that it may not come for a long time. The chances are that really bad strikes come tens of thousands of years apart, and perhaps before the next one comes, the space effect will have reached the point where there will be a "meteoroid watch" in near space as there is now an iceberg watch in the North Atlantic.

We might compound the disaster by imagining that the meteoroid strike is that of an antimatter object. Antimatter is composed of particles of a nature opposite to those composing ourselves and the Earth. Antimatter combines instantaneously with matter to give rise to a release of energy (including radioactive radiation) about a hundred times as great, size for size, as that of the warhead of a nuclear bomb. A small piece of antimatter would, therefore, do as much damage as a much larger piece of ordinary matter. (Some have speculated that the great 1908 Siberian strike was caused by a small bit of antimatter.)

Antimatter can exist and some astronomers suspect that there may be whole universes built of it or, within our own Universe, whole galaxies of it. However, it seems quite certain that our own Solar system, indeed our entire Galaxy, is made up of ordinary matter only. Any antimatter objects of any size would have had to wander in from other galaxies at the least and that is as unlikely, and as unpredictable, as a collision with a small black hole.

Suppose, then, we contract our view further and consider the Earth alone. Is there any chance it might, for some reason, explode, say, or that its axis may tip over?

Virtually none. The Earth has existed stably in pretty much its present form for over four billion years, and there is no reason it should not exist so for billions of years more if it is left to itself. As for the tipping of the axis, something called the law of conservation of angular momentum makes that so unlikely we need not concern ourselves with it.

The conservation of angular momentum does not, however, prevent some of the turning effect from being shifted from the Earth to the Moon. This means that the Moon is very slowly drifting away from us and the Earth's rotation is very gradually growing slower. As the day lengthens, temperature differences between day and night and between winter and summer would grow more extreme until Earth becomes an unfit abode for life. This change is so slow, however, that it will take many millions of years before the change will be significant.

The surface of the Earth moves, to be sure. It is made up of a number of large plates and some smaller ones, which slowly shift. Since the plates are in contact, pressure of one against another may crumple one or both plates and produce mountains; or cause one plate to pass under another and produce an ocean deep. In other places, plates move apart and hot material from below wells up.

As a result of these movements, continents can, in the long run, drift together to form a single huge land mass or, having formed one, break apart again. These enormous changes, however, take place so slowly that it is some millions of years before significant alterations in the position of the continents would be detected, and the changes are therefore not catastrophic.

As the boundaries of the plates, there are minor instabilities—ones that are insignificant on a planetary scale but very important on a human scale. It is along those boundaries that volcanoes are formed and that earthquakes take place.

Ordinarily, volcanic eruptions and earthquake tremors occur only at considerable intervals in any one part of the world, and while they can cause loss of life and destruction of property, the damage is usually local and temporary, and humankind has been living with such events all its existence. But what if there are some effects that can activate this instability and cause the poor Earth to shake badly all over while all the volcanoes let loose with a roar?

We don't know of anything that would cause this.

There are speculations that the Earth might be affected by the Solar wind (particles that shoot out of the Sun in all directions, streaming outward through the planetary system) and that the Solar wind is, in turn, possibly affected by tidal effects in the Sun that are caused by the planets. Some planetary configurations can cause unusually large tidal effects, leading to a sharp rise, possibly, in the solar wind. If some fault is on the point of yielding suddenly, to cause a disastrous earthquake, the changes brought on by the solar wind may just nudge it over the top so that, in the end, earthquakes and volcanic eruptions, too, may be caused by planetary configurations.

However, there are a great many if's in this conjecture and even if all the if's come to pass, the result would be a local and temporary disaster of a kind that would happen at some other time, perhaps not long after, even if the fault were not nudged.

What about catastrophes that affect the oceans and atmosphere of Earth rather than its solid body? What about changes in climate?

For instance, every quarter of a billion years, the Earth seems to undergo a period of recurrent ice ages, where water in large quantities freezes, then melts, and where huge glaciers advance for thousands of years then retreat for thousands of years. Scientists have speculated on the causes of these periods but have as yet reached no consensus. We are in such a stage now and, over the last million years, the glaciers have advanced and retreated four times.

Are the ice ages over? Perhaps not. The glaciers may someday advance a fifth time. We have survived the previous four advances, but human beings were then few in number and were tribal hunters who could move with the slow advance or retreat of the ice. Now we exist by the billions and are tied to the land by our farms, mines, and cities. A fifth advance would be a catastrophe.

However, the interval between the glacier advances (under natural conditions) is tens, even hundreds, of thousands of years, and by the time the glaciers come again we may have developed the kind of climate control that would prevent it.

Such climate control might also prevent the reverse possibility—that the Earth would grow a little warmer and that the glaciers that still exist in Antarctica and Greenland would melt. This would raise the sea level two hundred feet and drown all the rich and populated coasts of the world.

A more subtle danger involves the cosmic-ray particles that steadily bombard the Earth and that originate from the star explosions and other violent events here and there in the Universe. Those particles, electrically charged and highly energetic, are the source of both hope and danger. In smashing through living things, the cosmic rays produce mutations. Most mutations are harmful and can drive individual life forms and even whole species to death and extinction; but some are useful and these form the drive that keeps evolution going.

If the incidence of cosmic rays increase, then the rate of mutation will go up; and the preponderance of harmful mutations will drown the few good ones so that there will be what is called a "great dying," when whole groups of species suddenly disappear.

Much of the cosmic rays are warded off and turned aside by Earth's magnetic field, which thus keeps the incidence of particles to less-than-harmful levels. This magnetic field waxes and wanes irregularly for reasons we don't understand, however. At the present moment, the magnetic field is waning, and in a couple of thousand years it may pass through a period of centuries in which it is virtually zero.

At such a magnetic-field minimum, the incidence of cosmic-ray particles reaching Earth's surface will rise. If, at the same time, there are star explosions relatively near our Solar system, that might raise the cosmic-ray-particle density to unusual heights. It is this that might cause a great dying. The extinction, seventy million years ago, of the large and flourishing families of giant reptiles usually referred to as dinosaurs, might have been caused in this way, and who knows what will happen to Earth's ecological balance, and to us, if it happened again.

However, the combination of magnetic-field minimum and cosmic-ray maximum is very unlikely, and the chance of unusual destruction is small.

What of danger from other life forms? We no longer have any fear of lions, tigers, or any of the large predators or angry herbivores, but what of smaller animals? What of rats that grow ever more vicious and clever? What of insects that grow immune to insecticides? What of disease germs that may spread from human being to human being directly from, or by way of, insects and rats.

In the fourteenth century, the Black Death struck without warning and may have killed a third of all human beings alive in just a quarter century or so. This was the greatest catastrophe ever to strike humankind in recorded history. Many people then thought (and one can scarcely blame them) that the world was coming to an end.

There have been other plagues—of cholera, smallpox,

typhus fever, yellow fever—though none as deadly as the Black Death. As late as 1918, a worldwide influenza epidemic killed almost as many people as the Black Death did, though these represented a far smaller percentage of the world population than was true in the earlier case.

Might another vast epidemic arise and wreak incredible destruction? The answer, of course, is that such a plague, or a growth of vermin, could start at any time. It is hard to believe, however, that modern medical science could not deal with other life forms if fully mobilized for the purpose.

What, then, of *human* activities? Is humankind itself hastening the coming of any of these possible catastrophes, or making them worse, or even inventing new ones?

So far, nothing human beings can do will seriously affect the Universe, or any star, or the Sun or any of the planets, or even the body of the Earth itself. We can, however, affect Earth's atmosphere and we have been doing so.

Humankind has, for instance, been burning carbon-containing fuel—wood, coal, oil, gas—at a steadily accelerating rate. All these fuels form carbon dioxide, which is absorbed by plants and by the ocean, but not as fast as we form it. This means the carbon-dioxide content of the air is going up very slightly, Carbon dioxide retains heat and even a very slight rise means a slight warming of the Earth's temperature. This may result in the melting of the ice caps with unusual speed and before we have learned climate control.

In reverse, our industrial civilization is making our atmosphere dustier so that it reflects more of the sunlight and cools the Earth slightly—thus making it possible for a glacial advance in a few centuries, before we have learned climate control.

To be sure, the two effects seem to be nearly in balance now, and humankind is now making an effort to switch to non-fuel energy. Looming ahead of us is geothermal, hydroelectric, nuclear, and solar forms of energy and with these we may avoid the Scylla of melting ice caps and Charybdis of advancing glaciers.

Nuclear sources of energy can produce dangerous radiation, however. In particular, nuclear fission, which we are using now, not only offers a chance of core meltings that might liberate radioactivity over a large area, but constantly produces radioactive materials that are highly dangerous and that must be kept out of the environment for thousands of years.

The spreading use of nuclear-fission power plants keeps raising the nightmare of death by radioactive fallout for millions, of sections of the Earth turning radioactive through leakage of the stored ash, of the stealing of nuclear fuel by terrorists for use as the ultimate blackmail weapon. Many nuclear scientists, however, assure us that the dangers can be controlled and lived with.

Perhaps they are right, but an even better hope is that we will switch to nuclear fusion (not yet shown to be practical) which may lessen the danger of radiation considerably, or to solar energy, which should remove the danger altogether.

On the other hand, the nations may deliberately poison the Earth with radioactivity by using nuclear explosives in a vast war. (When the first nuclear bomb was exploded at Alamogordo in July, 1945, so little was known of nuclear reactions that some scientists feared that the chain reaction of exploding atoms might spread to the atmosphere and ocean and that an end to all life on Earth might follow one giant explosion that would virtually destroy the planet.)

Nuclear bombs have not, however, exploded the planet, and so far, world leaders, whatever their faults, have seemed to recognize that a nuclear war would leave no victors, few survivors, and a ruined planet. We may hope (rather wistfully, perhaps) that they will continue to understand this in the future.

The advance of science in other directions may involve catastrophic dangers. War weapons need not be nuclear bombs to lead to unimaginable destruction. The use of nerve gases, biological weapons, climate control, laser beam "death rays," and others are each more insidious and quiet, yet may, in the end, prove just as dangerous as nuclear bombs.

Even the advances of peacetime have their dangers. Advances in computer technology may lower the role of humanity and make human beings almost useless. Almost any technological advance may produce waste products that dangerously pollute the Earth. Chemical poisons fill the waters and the soil. Automobile exhaust and factory smoke fills the air. Pollution need not even be material. There can be noise pollution, light pollution, heat pollution, microwave pollution.

Everywhere, humanity's products beat upon the Earth which cannot, it seems, absorb it all. Even the noblest efforts of medicine may be harmful. So many individuals may be allowed to live through the help of advanced medical techniques, some maintain, that the "weak" and "unfit" will flourish, filling the human gene pool with undesirable genes whose catastrophic potentialities may someday make themselves felt.

Will we poison the Earth, kill the ocean, reduce the planet to one worldwide desert? We may not. There are ways of preventing pollution, of even reversing pollution, if humanity cares to take the trouble—and the expense.

A rather unusual route to possible catastrophe, revealed only recently, involves the ozone layer. About fifteen miles high in the atmosphere are small quantities of ozone (an energetic form of oxygen) which has the property of being opaque to ultraviolet light and of preventing most of the Sun's ultraviolet from reaching Earth's surface. It has been there ever since the Earth's atmosphere gained its free oxygen—that is, for half a billion years at least.

Human beings are now using spray cans in ever-increasing numbers where the sprays are driven out by very stable "chlorofluorocarbons" which emerge with the spray. These chlorofluorocarbons are very stable and remain in the atmosphere indefinitely. Eventually, some of it filters up to the ozone layer where, it is suspected, they may act to change the ozone molecules to ordinary oxygen.

If the ozone layer is destroyed in this way, floods of ultraviolet will reach the surface. Ultraviolet radiation is far less energetic than cosmic rays, but will reach us in far greater quantity. There might be a great dying, the extinction of many species that would greatly alter the planet's ecological balance. Through that, human beings would be gravely endangered even if they protected themselves from the direct harm by ultraviolet radiation.

Still, humanity is aware of this possible danger now and may take steps to prevent it.

Another subtle danger arises from recent microbiological experiments in which bacteria are having their genes altered and in which genes from one simple form of life are introduced into another. There is the possibility that some altered form of microorganism may be capable of causing some disease (cancer, for instance) against which the natural defenses of the body may not work. If such a microorganism escapes, it may be the Black Death all over again, or worse.

The chance of such an accident is admitted to be very small, but even a very small chance seems frightening and the people engaged in such work have voluntarily agreed to suspend such experiments until appropriate safety measures can be put into force.

This is an example of the way in which the dangers accompanying scientific advance can be foreseen and guarded against if people are willing to consider the nature of the advance thoughtfully and if they are willing to take appropriate countermeasures.

What, then, if nothing happens and humankind just continues to go on as always, without any significant catastrophe at all?

That, *too,* can be a catastrophe, and perhaps the worst.

Ever since *Homo sapiens* has been on Earth, his total numbers have increased from century to century. (The only exception was the Black Death century.) What's more, the increase has itself been proceeding at an increasing rate. As of 1976, the total world population is at the record high figure of 4 billion and the rate of increase is at a record high of 2 per cent a year, which means a doubling of the population in thirty-five years.

By 2010, then, if things continue as they are, the world population will be 8 billion. It doesn't seem likely that 4 billion mouths can be added to the present world population in only thirty-five years without widespread famine.

Under planetary famine conditions, the mad rush to extract food from the earth and the sea at all costs, and the drive to make use of any kind of energy, may permanently pollute and damage the Earth's ecological balance in ways a less desperate humanity would never countenance.

As the ill-fed and starving crowds multiply, the despairing attempt to hoard food or to steal from others will break down order and turn humanity into predators against each other. Any nations that retain a semblance of prosperity may, in desperation, press the nuclear button to force some sort of control over the rest of humanity. All in all, the pressures will cause the towering but rickety structure of civilization to collapse.

This is the catastrophe we must fear. All other possible catastrophes may come or may not come. If they come, they may not do so for millions or billions of years.

But if population is not controlled, civilization and most of humanity will face a catastrophe that will surely come, and within half a century. It is the population problem, then, that and nothing else, that should be the first order of business for humanity.

Afterword

I can't leave the public on a note of despair. Actually, it seems as though the general rate of world population increase reached its peak in the early 1970s.

It is now declining and may continue to decline. Partly this is due, alas, to an increase in the death rate in the poorer parts of the world, notably in south Asia, Partly, though, it is due to a decline in the birthrate, notably in China.

Public awareness of the problem of population is increasing, especially since growing shortages in energy and other resources seem inevitable for the near future. More and more governments are beginning to face the bitter facts.

There is no way of making a rapid about-face, of course, no way of escaping scot-free from the misery we are bringing on ourselves—but we are buying ourselves additional time. Things may turn out not to be as as bad as they might have been and perhaps with luck, endurance, and hard work, civilization may yet survive.

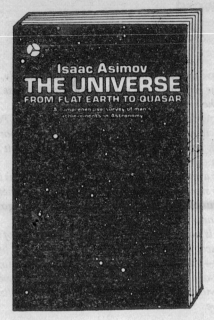